Franz Zeier

BOOKS, BOXES, AND PORTFOLIOS

BOOKS, BOXES, AND PORTFOLIOS

Binding, Construction, and Design
Step-by-Step

Franz Zeier

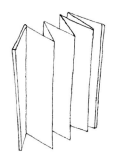

DESIGN PRESS

English translation by Ingrid Li
Design and illustrations by Franz Zeier
Typesetting by V & M Graphics, Inc.

Copyright © 1983 by Paul Haupt Berne
Translation copyright © 1990 by Design Press

10 9 8 7 6 5 4
Printed in Hong Kong
Reproduction or publication of the content in any manner, without express permission of the publisher, is prohibited. The publisher takes no responsibility for the use of any of the materials or methods described in this book, or for the products thereof.

Design Press offers posters and The Cropper, a device for cropping artwork, for sale. For information, contact Mail-order Department. Design Press books are available at special discounts for bulk purchases for sales promotions, fund raisers, or premiums. For details contact Special Sales Manager. Questions regarding the content of the book should be addressed to:

    Design Press
    11 West 19th Street
    New York, NY 10011

Design Press books are published by Design Press, an imprint of TAB BOOKS. TAB BOOKS is a division of McGraw-Hill, Inc. The Design Press logo is a trademark of TAB BOOKS.

Library of Congress Cataloging-in-Publication Data

Zeier, Franz.
    Books, boxes, and portfolios: binding, construction, and design step-by-step / Franz Zeier.
        p. cm.
    Includes bibliographical references.
    ISBN 0-8306-3483-5
      1. Bookbinding—Handbooks, manuals, etc.   2. Book boxes-
-Handbooks, manuals, etc.   3. Book covers—Handbooks, manuals, etc.
    I. Title.
    Z271.Z43   1990
    686.3—dc20   89-28644
                                                                     CIP

INTRODUCTION

Instructional texts on bookbinding generally assume that the reader already has some basic information of the kind that can be gained by an apprentice from a master in a direct and immediate way. The beginner often searches in vain for just what has been omitted. Other primers attempt to compress too much material into a little space, at the expense of the basics.

It seems useful and necessary, therefore, to focus on this humble but essential aspect of bookbinding and to do it accurately, in detail, and with a sense of urgency. My aim is also to put back into their rightful place the "simple" projects such as covered boxes, portfolios, and books, and to counteract the common notion that in the art of bookbinding, creativity starts with leather-bound volumes, with real — or artificial — cords and gold tooling.

Ideally a handbook such as this one should encourage the spirit of creativity. I have attempted to overcome the awkward division between technique and aesthetics. My book is intended to treat the material thoroughly and with respect, to offer technical information and stimulate innovative design at the same time.

For two decades I have taught future industrial arts teachers bookbinding and other paper crafts. I have had these teachers in mind in writing this book, but it will be useful to anyone who wants to learn about bookbinding as an educational pursuit, an enjoyable hobby, or both.

I have paid great attention to structuring this book clearly. To make it easy to use as a handbook and source of reference, key words and phrases from the text are called out in the left-hand margin. The drawings and color plates refer obviously to the topics of the illustrated paragraphs.

The projects are presented in order of increasing difficulty. Every step is explained carefully and should be easy to follow. On the other hand, the user of this book is not weighed down with irrelevant jargon and scientific information that has no bearing on the topics at hand.

Since drawings are more personal and often more to the point, I chose to use them rather than photographs. (All the drawings in this book were executed without the use of rulers and measuring equipment, except for the six painted polygons shown on page 129.) The color illustrations too are intended to provide more than just illustration. They are designed to share my personal view of the subject. They are not meant to be patterns: rather, they should serve as inspiration.

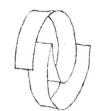

WORKING WITHOUT MACHINES

A special feature of this book should be pointed out: All the projects can be completed without the use of equipment like board shears, guillotine, and standing press. This should be a major advantage to most readers. There is, however, more than one positive effect of this restriction. To relinquish mechanical tools allows an even working pace and creates an atmosphere of concentration. Countless trips back and forth from machines are eliminated and replaced by procedures done by hand, often more quickly. This method should be appropriate to the educational goal of teachers. It also counters in a small but significant way the spreading blind faith in the powers of the almighty machine.

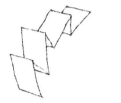

TO MY COLLEAGUES

No expert who thumbs through this book and wonders at the detailed descriptions will forget what group of readers I am addressing. I am well aware that most projects can be approached in various ways, and good results can come from different procedures, but it is important to start out with a tried and true method and develop variations when skill and experience increase.

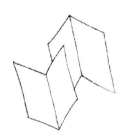

HISTORICAL BOOK FORMS

The chapter on historical book forms discusses examples of books and covers typical in various cultures and allows us to see the book that we are familiar with in a new and wider context. Practical applications and adaptations of these historical forms are encouraged, possibly for the first time.

GEOMETRIC SOLIDS IN PAPER

The inclusion of the chapter on making the geometric solids is only superficially surprising. If manual dexterity in dealing with paper is desired, the construction of multisurfaced figures serves this goal. The teaching of both geometry and art can be enriched by these projects. Part game, part work, they will reward with fresh insights anyone who tries them.

An interest in geometric forms can lead to a well-founded understanding of sculpture and architecture, of three-dimensional manifestations of our natural and artificial surroundings in general.

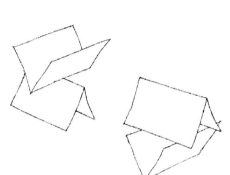

LIMITS

Earlier I explained why leather bindings, blind and gold tooling, gilt edges, and similar techniques are not discussed in this book. Any attempt to incorporate them into a school curriculum would be unwise — I even think it superfluous for hobbyists — because the choices and possibilities of preliminary techniques are so rich that this book was specifically drawn, painted, and written to explore them.

DECORATED PAPERS

Techniques involving decorated papers were not included for two reasons. Lately there has been an increase in literature on the topic, and anyone who wishes more information can easily obtain it. It is my intention to encourage the pursuit of the desired effect without the use of extensive color and decoration. More about this in following chapters.

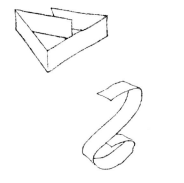

EXACT AND INEXACT WORK

Again and again in these pages I demand precision, but far be it for me to discourage those who are truly unable to work with a high degree of exactness. Creative freedom, however, should not be confused with sloppiness. I am convinced that even those who seem ill-equipped for manual disciplines can produce pleasing work if they trust in their abilities and put them to good use.

An object made with love and care can be a thing of beauty, a whole, even when it is imperfect; by contrast another one, although flawless, can leave us untouched and strike no sympathetic chord.

True amateurism should not be looked down upon. In times like ours any manifestation of liveliness must be cherished. We should ask ourselves if amateurism could not be this: to master a craft in a technical way only and to produce the most meager results in the field of creativity.

THE SPECIAL STATUS OF CRAFT

Obviously one cannot master a craft in a ten-hour course. Even learning how to ski or to smoke a pipe requires more time than that. To become proficient in a craft resembles the slow and steady process of learning a foreign language: step by step a strange new surrounding becomes familiar. But the joys of new discoveries by far outweigh the difficulties.

While I inveigh against a certain cuteness in crafts, an attitude of carelessness, I do not mean to advocate stern seriousness. A necessary element of playfulness will be addressed in coming chapters. Hard and determined work often brings gratifying results, while tinkering often leaves us unsatisfied.

Nothing good ever comes from work done without prudence, intensity, and indeed passion. If the right attitudes do not yet exist, they have to be developed. The result will be a feeling of wholeness that is so often missed in today's work and professions.

Is it not substance and meaning that we are longing for in our daily lives? The most wonderful result of our involvement in craft can be this: to stretch the boundaries of the ordinary and add a new colorful dimension to being alive.

ACKNOWLEDGMENTS

I would like to thank the following persons: Dr. Louise Gnädinger for her editorial help, Kurt Dinkelacker for his advice on technical problems, and Heinz Dieffenbacher for many a discussion on design and layout. My gratitude to Dr. Max Haupt for his unwavering goodwill and his conscientious care for this book. For the English-language edition, I thank Babette Gehnrich for her help with technical terms.

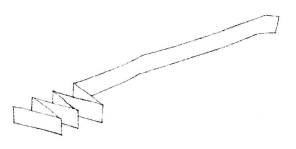

CONTENTS

THE FUNDAMENTALS   12

Materials   13
  Paper   13
  Paper Weight   16
  Cardboard   16
  A Paper Sample Scrapbook   17
  Handling Paper   20
  Flattening Paper   21
  Paper Formats   22
  Book Cloth   23
  Tools   24
  Truing A Triangle   26
  Weights   27
  Checklist of Tools   27
  The Work Space   28

Manual Techniques   30
  Folding and Creasing   30
  Creasing   35
  Scoring   37
  Tearing and Cutting   38
  Trimming A Sheet   42

Adhesives and Gluing   43
  Adhesives   43
  Adhesive Chart   46
  How To Glue Properly   47
  Mounting Paper on Cardboard   47
  General Hints   47
  Drying A Stack of Paper   55
  Edge Treatment   56
  Mounting Posters   61
  Mounting Paper on Paper   65
  Mounting Cardboard on Cardboard   66
  Stretching Paper   68
  Gluing Tissue Paper   69
  Indirect Gluing   70
  Gluing Book Cloth   71
  Gluing Unsized Cloth   72
  Gluing A Series of Sheets   73

THE PLATONIC SOLIDS   74

  General and Specific Properties   75
  Historical Facts   77
  Educational Uses of Paper Models   80
  Building A Paper Model   81
  The Tetrahedron   84

  The Cube   88
  The Octohedron   89
  The Dodecahedron   90
  The Icosahedron   91
  Large Paper Models   91
  The Archimedean Solids   92
  Inside A Cube   93

THE MAT   96

  Proportions   97
  The Folded Mat   99
  Mounting A Drawing in the Mat   100
  The Uncovered Two-Piece Mat   101
  The Covered Two-Piece Mat   103
  The Beveled Cut   103

BOXES   104

  A Folded Box   105
A Box with A Cover   108
  Construction   109
  Edge Reinforcement   114
  Covering the Box   117
  Lining the Box   118
  A Box Completely Covered with Paper
    or Cloth   120
  The Hinged Cover   123
  The Hinged Side   124
  Partitions   127
  Inserts   157
The Hinged Box   158
  A Box with Cloth Spine   159
  Construction   159
  Edge Reinforcements and Hinges   160
  A Hinged Box Covered with Cloth   161
A Round Box with A Cover   164
  Construction   165
  Covering the Round Box   169
  Lining the Round Box   171
A Round Box with A Lip   172
  Construction   172
  The Lip   173

PORTFOLIOS   175

  Creased and Scored Portfolios   177
  A Portfolio with Ties   180
  A Portfolio with A Cloth Spine   181

Corners   184
Ties   186
A Paper-covered Portfolio with A Cloth
   Spine   187
   Small Corner Reinforcements   191
   A Full-Cloth Portfolio   192
Portfolios with Flaps   194
   A Portfolio with A Side Flap   194
   Portfolios with Three Flaps   197

BOOK FORMS   202

The Scroll   204
The Palm-Leaf Book   207
The Accordion Book   211
   A Flexible Accordion Book   212
   An Accordion Book with Hard Covers   214
Side-Sewn Books   218
The Loose-Leaf Book   225
The Codex   230
Pamphlets   232
Adhesive-Bound Books   236
The Brochure   240
   Preparations   241
   Trimming   242
   Collating   243
   Sewing   244
   Gluing the Spine   247
   The Cover   248

THE HARDCOVER BOOK   250

Preparations   252
   Repairing A Book   252
   Reinforcing the Signatures   256
   Inserting Pictures   257
   Endpapers   258
   Trimming   261
The Book-Block   263
   Sewing on Tapes   263
   Swell   264
   The Thread   266
   Gluing the Spine   271
   Rounding the Spine   272
   Colored Edges   274
   The Headband   275
   Reinforcing the Spine   276
The Case   277
   Construction   277

Covering the Case   280
Connecting the Case and
   the Book-Block   282
The Paste-Down   283
Titling   284

PHOTOGRAPH ALBUMS   288

A Side-Sewn Photograph Album   289
An Adhesive-Bound Photograph Album with
   A Cover   293
A Sewn Photograph Album with
   A Cover   296

BIBLIOGRAPHY   301

SUPPLIERS   302

INDEX   303

THE FUNDAMENTALS

The steps described under this heading are not meant to be followed one by one. This approach would be too abstract and fruitless. The exercises should rather be incorporated in little projects: the construction of a simple portfolio, box, or geometric form. It seems obvious that a beginner might want to practice cutting and folding on scraps of paper. Photographs, posters, or pictures of all kinds can be used for first attempts at mounting. This chapter is mainly a collection of problems and solutions and should serve as a reference and guide throughout the work on later projects.

As a matter of course we think of wood, metal, and plastic when we talk about tools. The most remarkable tool of all, however, the one closest to us, is often forgotten: our own hand. It carries out any conceivable action, does it slowly or swiftly, vigorously or gently, with the highest degree of adaptability. Without delay it carries out our plans. We only have to imagine the diversity of tasks our hand is capable of performing to be amazed. During the process of sewing together sheets of paper it picks up delicate needles, guides the thread, positions the metal weights securely, counts the sheets of paper, holds and smooths them, reaches for the bone folder, the scissors, and the cake of wax. Our hands can manipulate objects rough and smooth, tiny as paper snips and large as lithography stones. Remarkable indeed.

The hand acts as tongs, pincers, hammer, container, shovel — and much more. There is always a connection to the arm, the whole body, and to the spirit. What is expressed through the spirit is expressed through the hand. A proverb of native Americans says the hand is the tool of the soul. It withers with the soul. Then, of course, it can no longer fulfill its duties.

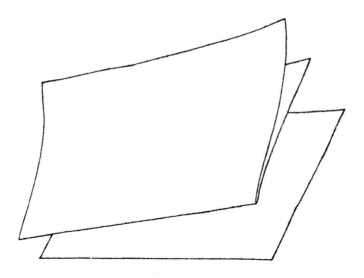

MATERIALS

Bookbinders use a wide variety of materials, but only the ones relevant to the projects in this book will be dealt with in any detail here. For more information on the origin, production, specific properties, and use of these materials, see the Bibliography.

PAPER

Grain	It is essential to understand that paper has a grain, the direction in which most of the fibers are aligned. As the pulp — consisting of 99 percent water and 1 percent fibrous materials, fillers, sizing, and pigments — is poured onto the moving screens of the paper-making machines, most of the fibers settle in the direction of movement. All industrially produced papers show this property, which causes wet paper to expand more in width than in length. This has to be taken into consideration when paper is to be glued, and it will be discussed in more detail in the appropriate chapter.
Expansion	

Short grain

Long grain

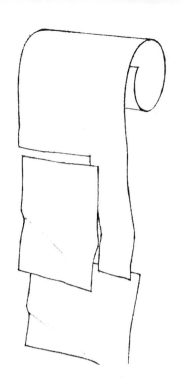

Sheets that are cut from the web (the sheet of paper coming from the paper-making machine) sideways are called short grain. Their grain direction runs parallel to the shorter side of the sheet. Sheets that are cut from the web lengthwise are called long grain. Their grain direction runs parallel to the longer side of the sheet.

If measurements of a sheet of paper are given, the number that designates the edge of the web is underlined. Thus, $\underline{24} \times 36$ means that you are dealing with a short grain paper.

Finding grain direction by folding

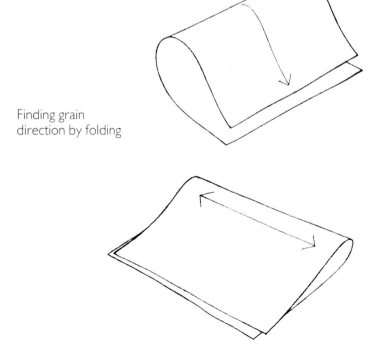

Different methods can be employed to determine the grain direction in any piece of paper that is at hand. The simplest one is to fold the sheet in both main directions and check the resistance of the paper. The paper will offer greater resistance to folding across the grain (and will tear less easily). In the illustration the arrows indicate the grain direction.

Moisture test for grain direction

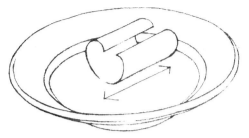

A 1½ × 1½ in (4 × 4 cm) square cut out of the sheet in question and dampened on one side only will readily indicate the grain direction. The paper curls with the direction of the grain as the axis of the curl.

Fingernail test for grain direction

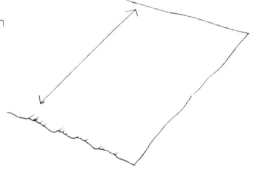

Using a third method, pull two adjacent edges between your index finger and your thumbnail. The cross-grain direction will appear more stretched and wavy.

The trained observer is often able to determine the grain direction of a piece of paper just by looking at it.

Fiber content

Papers containing wood and so-called acid-free papers both consist mainly of wood fibers, but acid-free papers are made of pulp that has been chemically treated. (One exception is 100-percent rag paper, which is made of cotton-fiber pulp.) Acid-free papers do not yellow, or only slightly.

Discoloration

Grammage
Basis weight

In most of the world, paper is described in terms of its weight in grams per square meter, or gm^2, while in the United States, weight is given in pounds per ream (500 sheets), cut to the industry standard for the specific paper grade.

The table on the next page shows the weight in pounds and gm^2 of common types of paper, as a general guide for choosing appropriate paper for the projects in this book.

PAPER WEIGHT*

Basis Weight Per Ream						Grams Per Square Meter (gm²)
17" × 22"	25" × 38"	20" × 26"	24" × 36"	25½" × 30½"	22½" × 28½"	
Writing Paper	Printing Papers	Cover	Board Newsprint Tissue	Index	Bristols	All Types of Paper
x	55		50			81
x	60		x			89
24	x		x			90
x	x		60			98
x	70		x			104
28	x	x	x			105
x	x	40	x			108
x	75	x	x			111
x	x	x	70			114
x	80	x	x			118
32	x	x	x	x		120
x	x	x	80	72		130
x	90	x	x	x		133
36	x	50	x	x		135
x	x	x	90	x		146
x	x	x	x	x	67	147
x	100	x	x	x	x	148
40	x	x	x	x	x	150
x	x	60	x	x	x	162
x	x	x	100	90	x	163
44	x	x	x	x	x	165
x	x	x	x	x	80	175
x	x	65	x	x	x	176
x	120	x	x	x	x	178
x	x	70	x	x	x	189
x	x	72	x	x	x	195
x	x	x	x	x	90	197
		x	x	110	x	199
		x	125	x	x	203
		80	x	x	x	216
		x	x	x	100	219
		x	x	125	x	226
		x	150	x	x	244
		x	x	140	x	253
		x	x	x	120	263
		100	x	x	x	270
		x	x	x	125	274
		x	175	x	x	285
		x	x	170	140	307
		x	200	x	x	325
		x	x	x	150	329
		x	x	x	160	351
		130	x	x	x	352
		x	x	x	175	384
		x	x	x	180	395
		x	x	220	x	398
		x	250	x	x	407
		x	x	x	200	439

CARDBOARD

Cardboard, more accurately called paperboard, is technically defined as paper above an agreed-upon weight or thickness. It is produced by wrapping a still-wet length of paper around a specially constructed cylinder until the desired thickness is achieved. Cardboard may be single or multi-ply. Another kind of cardboard is made by sandwiching several simultaneously produced layers of paper between two cylinders and pressing them together while they are still wet. Thicker sheets are produced by gluing together thinner ones. Technically, all sheets 0.012 in or more in thickness are paperboard, but for our purposes we shall refer to relatively heavy, thick, and less flexible sheets as cardboard.

In the United States, cardboard thickness is generally given in point-size or caliper, the thickness expressed in thousandths of an inch, generally between 10 thousandths and 125 thousandths of an inch. There is no standard sheet size.

We use cardboard to make boxes, book covers, and portfolios, as well as to protect our work surface. Cardboard can be bought in paper and art-material or hobby supply stores, and of course in stores that cater to bookbinders. Larger quantities are available from wholesalers.

*Adapted from The Dictionary of Paper, Fourth Edition (New York: American Paper Institute, 1980).

A PAPER SAMPLE SCRAPBOOK

The purpose of making a collection of paper samples is obvious: to learn how to distinguish among different papers. Before starting a particular project, say, a box, the owner of a sample book has a better idea of the possible choices of materials and their properties. Another bonus, no less important than these practical advantages, is the increased sensibility that comes from understanding different paper types. Eyes and fingertips are only partially responsible for our powers of perception, but to sharpen one sense usually enhances our sensitivity as a whole.

How to start

Variety

A useful collection of sample papers should feature the most frequently used kinds of paper. It is self-evident that a complete collection of papers produced in any given country would be as impossible to compile as it would be useless, since there would be tens of thousands of them. Nevertheless, it seems a good idea to start by saving every possible usable scrap of paper, since the material should generate excitement. Among this abundance there will be white and colored papers; smooth and rough ones; brittle, supple, transparent, opaque, matt, shiny, thick, and thin ones.

17

Classification

As the collector inspects and handles the various scraps and sheets and holds them up to the light, he or she will want to sort and classify them. This should not, however, turn into a cold scientific endeavor; it should be done with interest and understanding.

But where to start, what point of view to choose? A rather obvious system seems a grouping according to use. Some applications come to mind immediately: bookbinding, writing, printing, drawing, wrapping, painting, and many more such as making decorations, origami, and paper models. Other possible criteria for organization include weight, fiber, content, or any other quality.

Almost certainly there will be items that defy classification. Experts can be of help in a case like this. Keep empty pages ready to accept future additions to the collection.

I suggest the use of an accordion book (see page 211). Its construction requires very little skill, and it adapts to the thickness of the collected samples. Other types of bindings would require shims as in photo albums (see page 288).

Gluing the samples

The paper samples should not be glued in completely, but only along their left edges. This way they can easily be handled for inspection. How to go about this project is explained on page 73. Descriptions should be placed inconspicuously underneath each sample or next to it. If the source is known, it should be included.

Aesthetic quality

Today we can choose from among well and beautifully made papers, but regrettably also inferior ones. That category does not simply describe low quality, such as a paper that turns yellow quickly, but refers as well to products that pretend to be something they are not. No paper should be made to look like

Color

fabric or parchment. There is such a thing as beautiful wrapping paper, but there cannot be a beautiful parchment imitation. Imitations in general are to be rejected.

Papers in cheap and loud colors should never be used. Harmonic and pleasing color combinations are usually achieved with muted or broken hues. More so than we often assume, color involves emotions. Textiles of African or Indian origin, even though dull in color compared to chemically dyed fabrics, dazzle us with wonderful harmonies.

Years of observation and experience sharpen the senses. Studying superior works of art, making color sketches in the zoo, in botanical gardens, or from museum collections can enrich and expand our sensibilities.

Grouping the samples

Here are some suggestions for arranging and grouping paper samples:

Writing paper
Drawing paper
Book (printing) paper and coated, or glossy, printing paper
Wrapping paper
Handmade paper (only the real thing)
Chinese or Japanese paper (hand- or machine-made)
Household paper
Specially treated paper
   a) sized
   b) coated
   c) lined
Kraft paper
Decorated paper
Various kinds of cardboard (chip board or grey cardboard, bristol board, illustration board, cover board, mat board)

Larger samples, as of cardboard, can be cut into the same format as the sample book and kept in a separate envelope.

HANDLING PAPER

Care of paper

Paper is a delicate material: a little fold or tear can render a whole sheet unusable. Tender, loving care will pay off handsomely. It is unlikely that we will ever have to move around whole stacks of paper, but it is necessary to learn how to handle single sheets properly, to carry them from storage to workplace without harm.

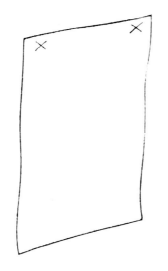

Medium-size sheets

A sheet of medium size should be held at the ends of an edge (the points are indicated by an x in the drawing). Larger sheets can be carried in one hand when you hold two parallel edges together.

Large sheets

Another possibility is to roll up a larger sheet. This should be done right on the stack from which it is taken. The roll can then be carried with one or both hands or under an arm, but it must never be squeezed together.

Even repeated handling should not mar a sheet of paper with any crease. These words of caution should not create timidity in handling paper, simply care.

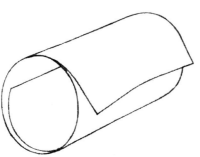

FLATTENING PAPER

Smoothing creases

Dampening the paper

A creased sheet can be smoothed in the following way. Apply clear water evenly and lightly with a sponge on one or both sides. Treat the whole surface, including the edges. Then place it between two layers of clean and absorbent cardboard and flat wooden boards. It should be weighted down and left undisturbed until it is completely dry, preferably overnight.

Certain papers should not be treated this way. Among them are very thin or brittle ones and any clay-coated paper. Watercolors, gouaches, hand-colored and otherwise delicate items can be smoothed by moistening the back side only. The cleaning of drawings and watercolors should be done only by experienced professionals.

Cleaning paper

Flattening curled sheets

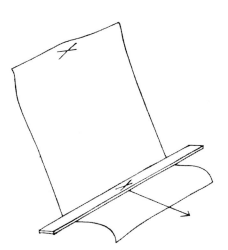

Obstinately curled sheets can be pulled over the edge of a table or under the edge of a ruler with more or less vigor as required. Look out for treated surfaces, thin sheets, or slightly ripped edges.

PAPER FORMATS

In the United States, paper is described by its basis weight, which is the weight in pounds of 500 sheets of that paper cut to a specified basic size. The table gives the most common basic sizes of the main types of paper.

U.S. standards

Paper Type	Basic Size in Inches
writing paper	17 × 22
book (printing)	25 × 38
cover	20 × 26
board, newsprint, tissue	24 × 36
index	25½ × 30½
bristol	22½ × 28½

International standards

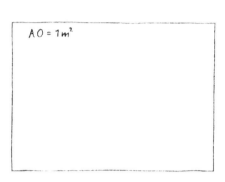

As mentioned before, most of the other countries in the world describe paper in grams per square meter, regardless of the type of paper or its size, and they have adopted a basic system of sizes developed by the International Standards Organization (ISO). The drawing illustrates the construction of the ISO A sizes used for cut sheets: the area of the largest A format (A0) is 1 square meter (though it is not a meter square). The next smaller ones are obtained by halving the previous sheet, so the ratio of the sides remains the same in all formats.

The table shows the A sizes in millimeters and inches.

	size in mm	size in inches
A0	841 × 1189	33.11 × 46.81
A1	594 × 841	23.39 × 33.11
A2	420 × 294	16.54 × 23.39
A3	297 × 420	11.69 × 16.54
A4	210 × 297	8.27 × 11.69
A5	148 × 210	5.83 × 8.27
A6	105 × 148	4.13 × 5.83
A7	74 × 105	2.91 × 4.13

cardboard
80 × 110 cm   31.50 × 43.31 in

BOOK CLOTH

Types of cloth

Selvage

Stretching and shrinkage

Sources

Referred to as bookbinder's linen in earlier times, but increasingly rare in days when even half- or quarter bindings are unknown to many, today's book cloth can consist of a variety of materials. Cloth made from natural fibers in various thicknesses can be used to cover books and boxes. One surface of the material is usually sized — treated to keep the glue from penetrating the fibers. Cotton or synthetic fibers sometimes have a layer of paper on one side and come in rolls; like paper they have a grain direction, which can be recognized at the selvage. The material can be stretched crosswise.

It is difficult to predict how a fabric will react when glue is applied or if it gets damp. It usually stretches less in length than in width and shrinks accordingly during the drying process. If untreated fabrics are used for bookbinding purposes, the instructions that are given in the chapter about gluing should be followed.

Small quantities of book cloth must be purchased from professional bookbinders or from hobby supply stores. Only larger amounts are available from wholesalers that cater to the trade.

Many available fabrics should be rejected for aesthetic reasons. Loud colors, iridescent surfaces, rough fabrics, and similar oddities, along with imitation materials, should never be used.

Special fabrics

There are special fabrics that are not used for covering. They include shirting, a thin, tightly woven white fabric of cotton or linen used for hinges and reinforcements, and lining strips — usually cotton gauze in a natural color — which are glued to the spine. Other special materials are headbands and thread, which is used to sew the signatures together and connect the signature to the tapes.

Headbands

Thread

Tapes

TOOLS

The following tools should be available to every bookbinder: a bone folder, an all-purpose or bookbinder's knife, scissors, a ruler, compasses, two carpenter's squares, an awl, a chisel, and a hammer.

Bone folder

The bone folder should not be large and awkward. It should fit the hand comfortably: 6 in is average size. Its shape should be that of a slender boat and can be altered with sandpaper. Its edge must be sharp enough to leave a clear crease, but never rough or honed to the point where it actually cuts the paper.

Knives

Knives have to be as easy to handle as bone folders. The blade must be securely attached to the handle, thin and absolutely sharp. Every bookbinder is well advised to learn how to sharpen his or her own knife. Since a dull knife can destroy the work of many hours in a matter of seconds, never hesitate to interrupt the work process to right the matter.

Scissors

Scissors should be of a good size and should cut precisely, even close to the tips.

Pencils

Choose a pencil that is hard enough to produce a thin, clear line.

Compasses

Compasses should be at least 6 in long.

Rulers

A ruler should be checked for accurate markings and intact edges.

Squares

A carpenter's square (45 degrees) and a 30-degree triangle are useful and should be checked for accuracy, particularly if they are made of wood. How to check a triangle and correct for deviations is explained on the following pages.

Sandpaper

Sandpaper files are often used and easy to make. See page 66 for instructions.

Care of tools

Caring for one's tools is an important aspect of the work, and nothing that is connected with it should escape your attention. The light source for your surface, the state of the glue pot, or the cardboard mat under your work— everything deserves care and interest. Because tools should function as extensions of the hand, never borrow someone else's. This is particularly true for bone folders and knives.

TRUING A TRIANGLE

Checking for accuracy

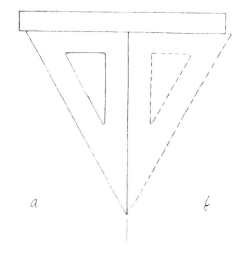

Follow this procedure to check the accuracy of your triangle. On a paper surface position a ruler and weigh it down. Position the short side of the triangle against it. Draw a fine pencil line along the side that forms a right angle with the ruler. Flip your triangle into a mirror-image position on the other side of your pencil line. Take great care not to shift the ruler during this move. If the pencil line still matches the side of the triangle, then its angles are accurate. Of course, the edges of the triangle should be checked for smoothness.

Adjusting an angle

Angles can be changed by gently moving the triangle across a large sheet of sandpaper. During this process the results should be checked frequently, using the same procedure as described above.

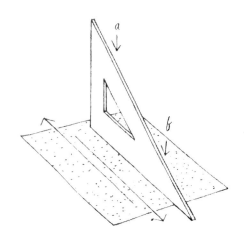

WEIGHTS

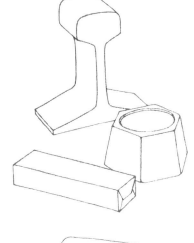

Metal weights

This book assumes no access to a bookbinder's presses; we rely on the use of weights. Various items can be used: odd chunks of metal, actual weights, or sections of railroad track. Two kinds are required: some lighter pieces, about 2 lbs each, for preparatory work and for sewing, and heavy ones, weighing 10 to 20 lbs, for pressing.

Sharp edges should be rounded to protect wood and paper that come into contact with these weights. To be safe, wrap the weights in paper.

Lithography stones

If you can come by them, lithography stones are excellent tools for the bookbinder. They are perfectly shaped as weights and work surfaces.

CHECKLIST OF TOOLS

Bone folder, about 6 in
Knife with one cutting edge, all-purpose or one made especially for bookbinders
Scissors (if possible, ones made especially for bookbinders)
Pencil (#2 or #3) and eraser
Rulers, one short (about 12 in) and one long (about 24 to 30 in)
Compasses
Awl
Medium-size hammer
Chisel
Boards made of pressed wood or similar material in dimensions of about 7 × 10, 10 × 12, and 15 × 20 in (two of each)

Weights wrapped in paper
Sandpaper files
Sewing threads of different thicknesses
Sewing needle
Tape (cotton), about ½ in wide
Beeswax cakes
Round brushes with diameters of about 1, 1½, and 2 in (two each)
Flat brushes, about 1, 1½, and 2 in wide (three each)
Three bowls, diameter about 5 in

THE WORK SPACE

Table

The first requirement of a desk or a table is stability. A solid table is therefore preferable to a makeshift one. It should measure at least 3 × 5 ft. The surface should be smooth and covered with cardboard about $1/8$ in thick. This protects the tabletop and allows paper cutting right at the working surface.

Protecting the work surface

The work area should always be kept in order since chaotic surroundings make fruitful progress difficult, if not impossible.

Keeping order

Position scrap paper, glue, and containers for brushes on the right-hand side near the area reserved for gluing operations. Measuring equipment and other tools are kept on the other side of the table. Some procedures are better performed while standing up, and for these a higher work surface is helpful.

Lighting

Artificial light

Good, but not even, lighting creates the most comfortable conditions. Light that hits the surface from an angle lets us see objects and their shadows and is preferable to a single overhead light source that surrounds the area. Direct sunlight on the work surface is undesirable, since it may cause paper and cardboard to warp and glued areas to dry too rapidly. The most adaptable light source is an adjustable lamp that can be positioned according to need. In addition, an overhead light can be useful.

Two light sources of equal strength cast double shadows and confuse the appearance of objects. Diffuse light such as that produced by fluorescent bulbs is completely unacceptable for our purposes, since it creates almost no shadows and robs objects of their familiar appearance, which diminishes the joy of the creative process. We are not just after the end product — our minds and our senses should be nourished along the way.

MANUAL TECHNIQUES

FOLDING AND CREASING

Paper folding, and what results from it, should not be approached in an abstract way. How does one fold a letter? Carefully or in a hurry? Coolly, with precision, or eagerly, with flying fingers that cannot help crumpling the paper? A folded sheet, closed and reopened, is a remarkable object at any rate. The subject has been painted again and again by artists such as Manet, El Greco, Vermeer, Holbein, and contemporary masters, and with good reason: It is a bright accent, it defines planes, and it offers stark contrasts with its surroundings.

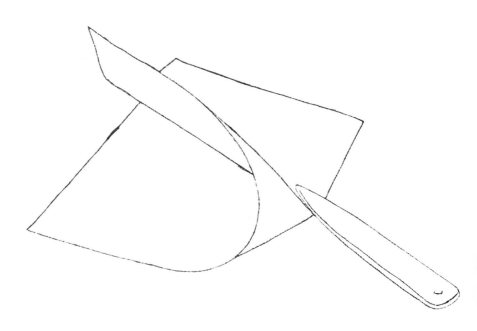

Letters, folded or unfolded, have been painted not only as poetic additions but as visual elements in their own right.

Have you ever watched a man paste a huge poster into place? A most impressive display of skill with glue and paper.

You can fold paper for a functional or a playful reason. Folding a letter would be the former, making a paper boat the latter. This chapter deals with functional procedures.

Folds and creases

Folds and creases are very much the same thing. A fold is produced without the aid of tools, as with a letter or paper boat. Creases are made with tools and appear sharper.

A folded sheet can be reopened, but the line along which it was folded remains forever visible as a sort of wound inflicted on the material of the fibers. Light and shadow make it recognizable, even when the sheet is smoothed out.

Folding and grain direction

A fold with the grain is easier to make and appears cleaner than one across the grain. The resistance felt in the latter case can make folding a heavy sheet impossible unless it is done in the right direction. Should folding across the grain be necessary, the paper has to be creased first.

Parallel fold

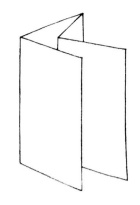

The drawing shows an example of frequently used parallel folds.

Accordion fold

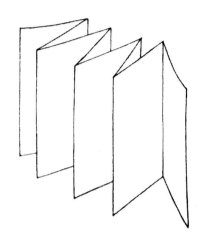

Another parallel fold, this time in zigzag form, is called an accordion fold and is described in more detail on pages 211 to 216.

Gate fold

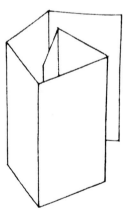

Gate folds have all creases in the same direction.

Cross folds

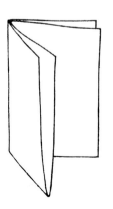

Cross folds, or broadsides, are mainly used for books and magazines and are named for the pattern that is formed by the creases.

Mixed folds

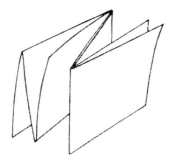

Mixed folds are a combination of parallel and cross folds.

Original sheet

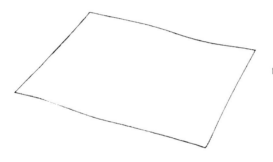

For a book, the grain direction must be parallel with the last fold.

Folio
One fold

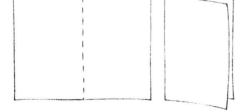
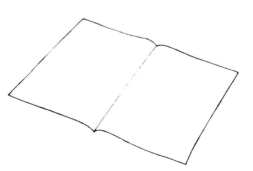

The sheet is folded once, and this is called a folio.

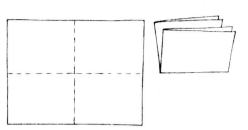

Quarto
Two folds

The sheet is folded a second time, at a right angle to the first fold. This is called a quarto.

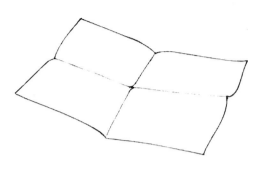

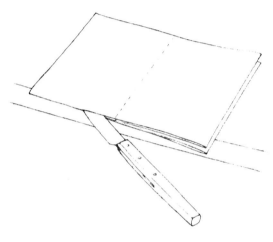

Before the third fold is made, the second fold should be cut about two-thirds of the way with a paper knife to avoid wrinkles that would otherwise be caused by the third fold.

Octavo
Three folds

The third fold is made at a right angle to the second fold and divides the sheet into eight sections. This is a three-fold or octavo. Most signatures (the folded-up sheet of paper that forms a section of a book) are produced this way. The three-fold signature produces sixteen pages.

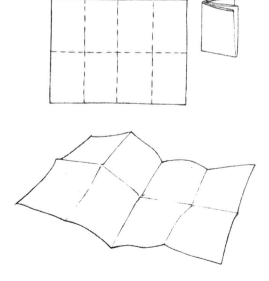

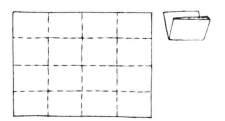

Four folds

If a fourth fold is made, cut along the third fold. The fourth fold divides the sheet into sixteen sections. It is often used for books printed on lightweight paper. It produces a thirty-two-page signature.

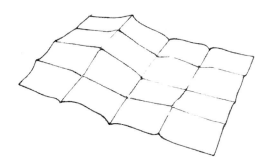

CREASING

How to hold the bone folder

When to crease

To hold a bone folder properly, rest your index finger on its upper edge while your thumb on the left and middle finger on the right side guide it, usually with no more pressure than you would apply to a pencil. The work should be done standing up, and the direction of the crease is not sideways but pointed toward yourself.

Creases are needed whenever a material is too thick to be folded by hand. A crease weakens the paper along the folding line and eliminates resistance. Since lightweight materials require thin creases and heavier papers wider creases, it is useful to have two bone folders of different widths.

Effect of the surface

Waxing

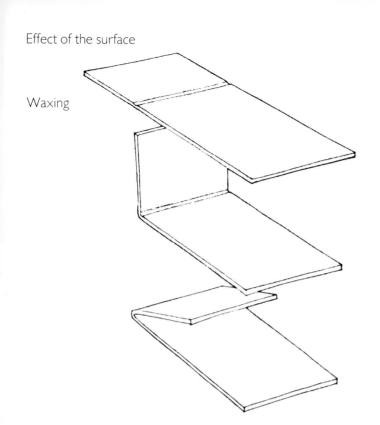

Working on a soft surface like cardboard will produce more pronounced creases. Lightly waxing the sheet will make creasing easier and protect the paper at the same time. Do not apply the wax directly, which would produce uneven results or shiny spots. Rather, transfer it with a clean rag or some cotton. Another method of waxing is a light touch with fingers that have been pulled through your hair.

Creasing cardboard

Even cardboard can be creased by hand with a little preparation. Use a work surface of two pieces of cardboard, with a small gap between them, glued on a third. The cardboard piece that needs to be bent is forced into the gap with the wide end of the bone folder.

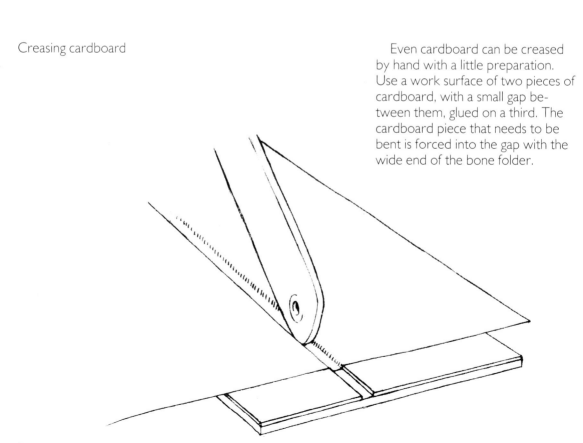

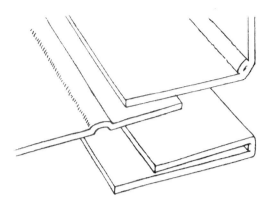

This is not a last resort but a legitimate technique, and produces a first-rate crease. You can hold the sheet in place with a ruler rather than your hand, but the bone folder should never touch the ruler. For cardboard more than about 1/32 in (0.5 mm) thick, you must use a machine called a creaser.

SCORING

When to score

If cardboard is so thick that it cannot be folded properly even after it has been creased, it must be scored. A score diminishes the resistance of the material against bending. This of course weakens the cardboard, so the score should never be deeper than absolutely necessary.

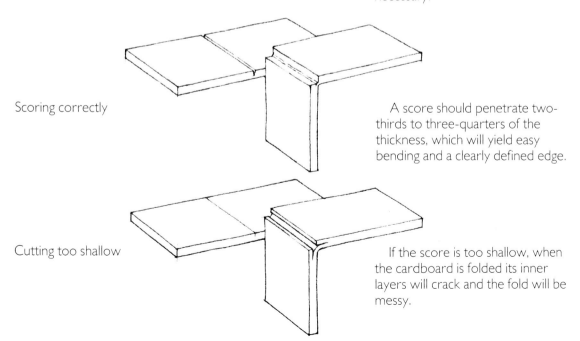

Scoring correctly

A score should penetrate two-thirds to three-quarters of the thickness, which will yield easy bending and a clearly defined edge.

Cutting too shallow

If the score is too shallow, when the cardboard is folded its inner layers will crack and the fold will be messy.

Cutting too deep

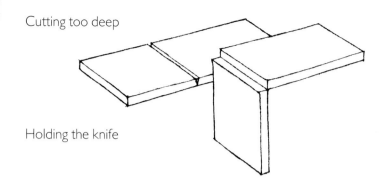

If the score is too deep, the remaining hinge will be too weak to support the fold.

Holding the knife

The cutting knife should be held at a steep angle; with heavy cardboard, almost vertically. Close your whole fist around the handle. That the knife has to be sharp is obvious.

Cutting and grain direction

As always, the grain direction must be reckoned with. The number of strokes just right for a cut against the grain might sever the parts from each other when you are scoring with the grain.

Smoothing rough edges

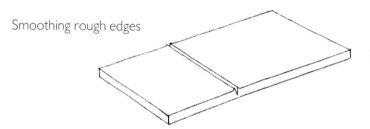

The raised edges on both sides of the cut can be smoothed with the help of the bone folder.

TEARING AND CUTTING

Separating by hand

To separate two halves of a sheet of paper you can simply rip them apart, but if you want the line to be straight you first must make a sharp fold and then tear along the fold.

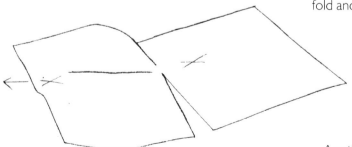

Another method, which everyone is familiar with, is to fold the paper, insert a knife into the fold, and separate the halves. A dull blade or a letter opener are of bet-

Separating by folding and cutting

ter use here than a sharp knife, which would have to be guided exactly horizontally to avoid damage, while the fold of the sheet is positioned flush with the edge of the table. Both methods produce more or less ragged edges.

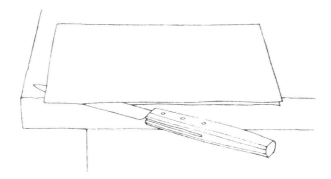

Cutting with scissors

Scissors are useful for cutting curves, for the short incisions that are needed when covering round boxes, and for longer cuts if it seems more convenient. Small deviations from the intended line are usually without consequence. Take advantage of the freedom of handling scissors instead of knives and rulers as often as possible.

Cutting with a knife

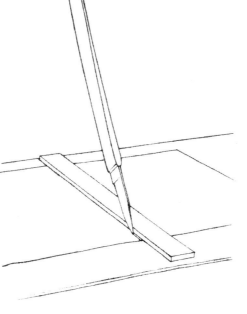

The most frequently used cutting technique employs knife and ruler.

The position of the knife and the pressure applied should not change during the cutting action. Always stand upright and guide the knife toward yourself, never sideways.

Cutting thin paper

Cutting thick cardboard

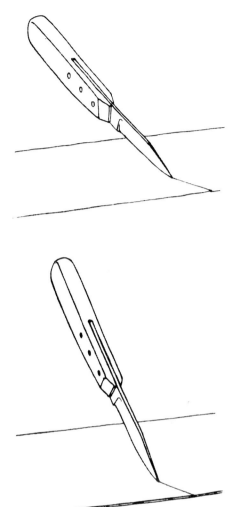

When cutting very thin papers, "drag" the blade across the surface, that is, at a small angle to the paper, like a bone folder. The angle increases with the thickness of the material, until it reaches almost 90 degrees for cuts through 1/8 in cardboard. In this case, enclose the handle of the knife in your fist.

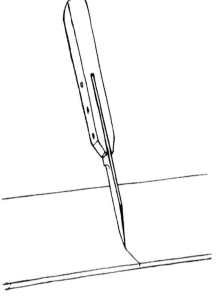

The first cut

Long cuts

Cardboard usually cannot be cut in one move. Create a guideline by making an initial shallow cut on the surface; into this place subsequent, increasingly deeper cuts.

Long cuts through thick cardboard are particularly challenging. If no helper is available to hold the ruler in place, clamp it to the tabletop on both ends.

Old cutting marks

Cutting whole sheets

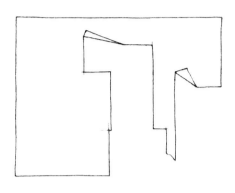

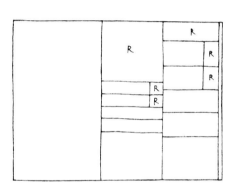

Saving scraps

Cutting surely

One danger lurks underneath the paper that you intend to cut: scores left by previous cuts, which could easily guide the knife in the wrong direction. Always check the cutting surface, and place new cuts deliberately left or right of existing marks.

The sequence in which pieces are cut out of a sheet deserves attention, not just because planning obviously limits waste but because sheets with many corners sticking out get entangled like fishing hooks in a drawer or folder. They get dog-eared and torn, and even endanger the smoothness of fresh sheets in the folder.

The drawing shows how pieces for a box with a cover can be cut from a sheet. First the rough edge is removed, then the two wide strips are cut from the sheet and separated into smaller units. The second strip contains pieces for the box cover.

The sections marked "R" are remnant materials. These are rectangles of varying sizes, all fully usable. The pieces should be separated by type of material and kept in different folders.

Even the smallest scraps should never be tossed out without a second thought. Collections of white and colored snips and scraps can be quite inspiring and provide hours of pleasure for a child.

Confident cutting requires practice. Do not stare at the tip of the knife or cramp your hand around the handle in desperation. As for the novice bicyclist who is advised not to look at the handlebar but at the road ahead, the secret of success lies in smooth movements.

TRIMMING A SHEET

First trim

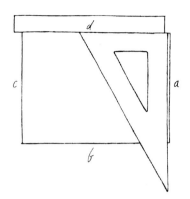

Second trim

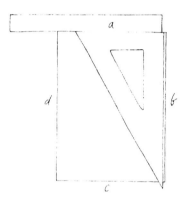

Third trim

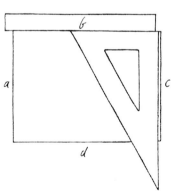

Fourth trim

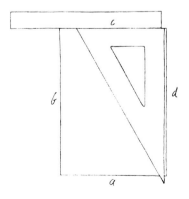

You usually need a sheet with exact 90-degree angles at the four corners. It is a common misconception that paper as it is purchased meets this requirement. In addition, some sheets come with ragged (or deckle) edges that are unusable in most cases for our purposes.

Often only a corner cut is needed; a trimming along two neighboring edges. Perfectly aligned measuring devices are necessary to detect irregularities in paper measurements and angles. Slight changes can be made on wooden tools as described earlier, but of course not on metal ones.

To trim a sheet of paper properly, follow this procedure:

First trim: Align paper and triangle with ruler and trim 1/16 in (1–2 mm) off side a. Cut only what seems necessary.

Second trim: Align the edge that was just trimmed with the ruler and repeat the above along side b.

Third trim: Again position the trimmed edge (b) against the ruler and trim along side c.

Fourth trim: Place side c against the ruler and trim the last edge.

ADHESIVES AND GLUING

ADHESIVES

No other profession relies on adhesives as heavily as bookbinders do. Amateurs often fear working with adhesives, but take heart: The job can be done and even enjoyed. Applying paste with free strokes onto a large sheet of paper can be positively invigorating.

We use two kinds of adhesives, paste and PVA, or a mixture of both. Today PVA takes the place of the animal glues that were traditionally used in the trade.

Recipe for paste

Pastes are derived from plants and can be prepared at home from wheat starch. Mix 4 tablespoons of wheat starch with about ½ cup of cold water. Stir until all lumps dissolve. Continue stirring vigorously and add about 2 cups of boiling water to form a thick mixture.

Preserving paste

While this is cooling, cover it with wet newspaper. The paste should look neither milky nor glassy. You can add a drop of Thymol (which comes in crystals that must be dissolved in alcohol) or refrigerate the paste to keep it from turning sour.

Lumps in paste

The mixture must be completely cooled before it can be used. Occasionally lumps will form in spite of careful handling. In this case strain the paste through cheesecloth into another container.

The following list of applications is necessarily incomplete; personal experience should be a guide to other uses.

When to use paste	Use paste:
	    to mount large pieces,
	    to mount lightweight paper,
	    to reinforce folds,
	    to line boxes and portfolios, and
	    to patch damaged spots.

Advantages of paste	A major advantage of working with paste is that stains can usually be removed easily with clean water. Sheets mounted with paste can generally be removed without damage.
PVA	PVA is sold under various tradenames. It can be thinned with water, but once dry, it becomes water-insoluble.
Care of brushes	Brushes encrusted with dried PVA cannot be saved. It is essential that brushes be cleaned in water after use, or that bristles be kept covered with water if the brush is to be reused soon.
When to use PVA	PVA is used to glue cardboard together, to attach heavy paper to cardboard, and to glue small areas.
Mixing paste and PVA	Mixtures of paste and PVA result in a very versatile medium. Even a small addition of PVA makes the adhesive water-insoluble after it has dried. The drying process itself is accelerated, and the mixture causes the paper to buckle less than it would with the use of pure paste.

On the other hand, PVA can be improved by the addition of a small amount of paste. It will dry more slowly, which can be a definite advantage, and it is easier to apply. A sheet that has just been mounted can be taken off if not too much time has passed, and paste can be used as a thinner instead of water during the process.

Moisture content

Both paste and PVA contain water; the latter considerably less. Paper and cardboard expand when they become wet and contract during drying, as we have already seen. We aim to use as little water as possible to avoid such undesirable side effects as buckled surfaces.

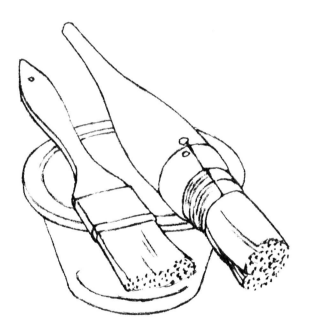

Contact cement

Adhesives, such as contact cement, that do not contain water do exist, but they cannot be used for our purposes. They should be reserved for temporary uses such as quick mountings for exhibitions.

Which adhesive to use for a given project is largely determined by one's own experience and preference. The binding properties, the resulting tensions, and working speed have to be balanced against each other in every case.

The following chart offers some guidelines:

ADHESIVE CHART

Paste	PVA	Use
100%	—	Lightweight paper on paper or cardboard
4 parts	1 part	Medium-weight paper on paper or cardboard
1 part	1 part	Heavy paper and woven materials on cardboard
1 part	4 parts	Lightweight cardboard on cardboard or wood
—	100%	Cardboard on cardboard

These suggestions have to be adapted depending on the size of the material to be glued. The same paper can offer a great deal of resistance in a poster size, while it might be easy to handle with smaller dimensions. The larger the sheet, the higher percentage of paste is called for in the adhesive mixture.

HOW TO GLUE PROPERLY

Gluing methods

The examples in the following paragraphs are typical of the many situations that can arise when you work with paper. To concentrate on the most important procedures is next to impossible, since almost every situation constitutes a special one. I try to be as concrete as I can and to leave adaptations to the reader. Further hints about adhesives and their uses are also to be found in some other sections of this book.

Brushes

Always have at your disposal at least two round brushes with a diameter of 1 ½ or 2 in and two flat ones about ½ in wide. All of them should be of good quality, made of natural bristles and in good condition. No bristles should ever come loose during pasting.

MOUNTING PAPER ON CARDBOARD

GENERAL HINTS

To understand the basic behavior of adhesives prepare a thin piece of cardboard, about 15 x 25 in, and a piece of paper of the same size.

Preparing the work surface

It is very important to work on a flat and clean surface. Protect the work surface with a single sheet of newsprint, never a whole stack, since folds and creases in lower layers could be overlooked and could later interfere. After one use, the sheet should be discarded. If a larger piece is needed, connect smaller sheets to each other with a dab of adhesive.

Applying adhesive

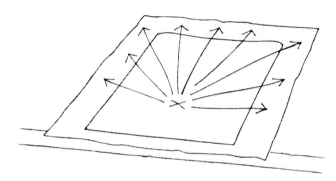

Put the paper to be mounted on a larger sheet of newsprint and apply paste in two steps. The round brush, unlike watercolor brushes, is not held between fingers but in the fist. Two fingers of the other hand hold the sheet at the lower edge, while the paste is applied in a pattern of radiating strokes towards and past the three edges.

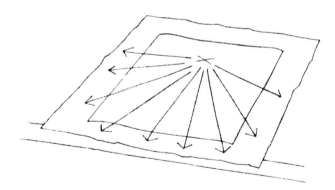

When the paste is evenly spread, move your hand to the upper edge of the paper and repeat the process in the opposite direction.

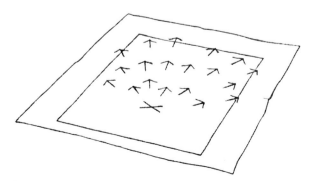

Some experienced bookbinders prefer to apply the adhesive in dots, especially at the beginning of the process, and they gently spread the adhesive in the last step.

Curling edges

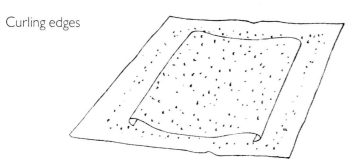

It soon becomes obvious that the sheet curls along the edges as a result of the stretching against the grain. It reacts as if it were alive.

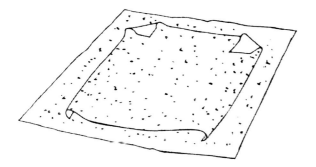

There are different ways to counter this effect:
   a. Pull the brush across the curling edges vigorously and repeatedly, until the material "tires" and stays flat.
   b. In particularly challenging situations, fold over the corners of the sheet as shown.
   c. In rare cases dampen the underlying paper.
   d. Apply the adhesive indirectly, as described on page 70.

Before you lift the sheet off, check for loose particles or bristles and remove them with the tip of a knife if necessary. Also make sure that the whole sheet has been covered with adhesive and fill any dry spots.

Lifting the sheet

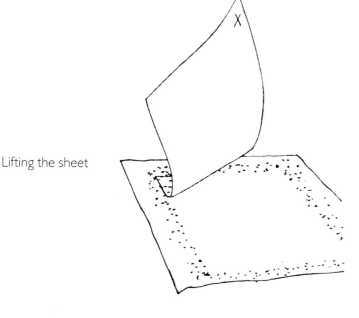

Gently pick up the near right-hand corner with your left hand and lift the paper so that it hangs downward. With your right hand pick up the diagonally opposite corner, so that the pasted side faces away from you.

Setting down
the sheet

Put the corner that you are holding with your right hand down onto the cardboard, matching its bottom right corner. With the left hand, lower the rest of the sheet onto the cardboard until one paper edge is flush with one cardboard edge, preferably in the same grain direction.

Rubbing down

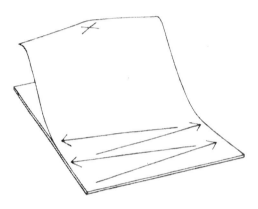

Rub gently and move your left hand to the middle of the opposite edge. With successive strokes of the right hand press the paper against the cardboard, while the left hand holds up the still unattached portion.

If the paper was covered evenly with adhesive and was allowed to expand, there should be no bubbles or creases.

Setting down
larger sheets

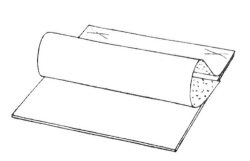

To mount sheets of medium size (up to 15 × 25 in), follow this method: Hold the sheet to which the adhesive has been applied by the two corners along one edge. Use both hands. Set the edge down on the corresponding edge of the cardboard, but roll the rest of the sheet until its dry side rests on the cardboard. Rub the initial edge, then unroll the rest of the sheet and proceed as with a smaller sheet.

Using a protective sheet

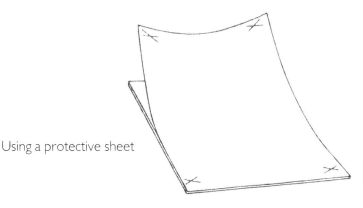

Large sheets require the cooperation of two pairs of hands. Cover the mounted paper with a sheet of smooth paper, and start rubbing it down, outwards from the middle. Lift the paper occasionally and check for air bubbles or trapped particles.

Rubbing down

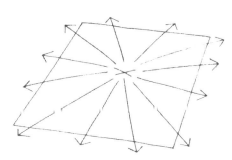

A word of caution: To rub down the sheet, use your whole hand but not your knuckles. Long and deep grooves in a finished mounted piece often bear witness to this faulty procedure or to an excessively heavy application of adhesive.

Why a sheet could come loose

When pure paste is being used it is especially important that you rub the sheet carefully. A common scenario is this: Everything seems nicely glued in the beginning, but after a week, a month, or even a year the paper separates from its carrier. There are two possible reasons: The application of paste was too thin, or too much time elapsed between applying the paste and setting the piece down.

Apply adhesive to paper, not cardboard

Always put adhesive on the paper, never on the cardboard. Applying adhesive to the cardboard will cause a myriad of waves and buckling to appear, and no amount of rubbing will counteract this. Try it!

Special situations

If occasionally the paper has to be rubbed down with a greater pressure, use the flat edge of the bone folder. Special care is required, because too much pressure can force the adhesive to spread unevenly, resulting in an uneven surface.

Mistakes

If adhesive oozes out around the edges:
    a. Too much adhesive was used;
    b. Too much pressure was applied;
    c. The adhesive was too thin.

Treating one side only

Push

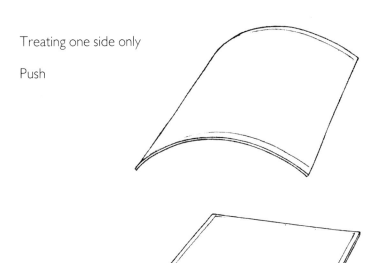

A novice might think that the work is now done. Let us see what happens if we stop here. The freshly mounted piece will arch, the side that carries the paper becomes convex, flattens after a while, and finally turns more or less concave.

Pull

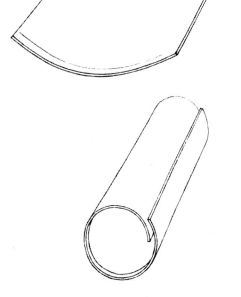

If you repeat the process with several layers of paper, a tube will form, provided that the grain directions are identical.

Counteracting tension

Two more procedures are essential to explain how to counteract tensions and how to dry a piece properly. To keep a mounted piece permanently flat you have to balance the tension forces acting on it. This means that a paper of very similar properties has to be glued on the other side of the cardboard, preferably at the same time and with similar intervals between steps. Whether the right balance was reached can only be determined when the work is finished.

Drying

The importance of the drying process is generally underestimated. Perfectly executed work can be ruined if the piece is left to dry in the open or not long enough. A piece that contains a high degree of moisture should not be put under pressure immediately. Instead it should be exposed to air, maybe leaning against a wall for about 15 minutes, to allow some of the moisture to escape.

Drying in open air

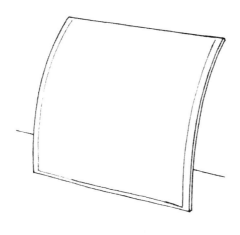

Weighting

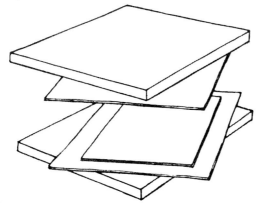

The cardboard, covered with paper mounted on both sides, is now placed between two absorbent sheets of cardboard of slightly larger size, and finally between two boards. Then a weight is put on top of it. The drying process takes at least 4 hours, during which the absorbent sheets of cardboard should be changed at least once, after the first ½ hour.

Drying time

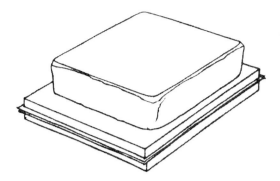

Just because a piece feels dry to the touch does not mean it actually is dry. Play it safe — allow some extra time.

Never use sheets of metal or plastic instead of wooden boards, to ensure that the moisture will not be trapped inside.

Using paper as blotters

To dry a piece between layers of paper is to invite disaster. Paper would absorb excess moisture, but it would buckle and probably ruin all the layers that are involved.

Using a press

Presses are not necessary for the projects in this book. Many a good piece of work has been destroyed by too much pressure. Be prudent; some metal weights or stones will usually suffice. Remember, however, never to put them directly on your piece.

DRYING A STACK OF PAPER

If many sheets of papers have to be dried at the same time, they should be stacked in alternation with sheets of cardboard. Think of the resulting pile as a wet paper pad. If you put it under pressure and leave it to dry from the outside in, all the cardboard and paper will contract and warp at the same time, with no hope of ever becoming flat again. To avoid this disaster, replace the damp cardboard layers with dry ones at intervals of 30 minutes at the beginning to as much as 5 hours at the end of the process.

These are some properties and conditions of working with water-based adhesives. Others will be explained later.

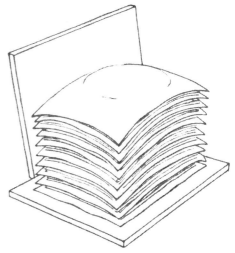

EDGE TREATMENT

It is essential to know how to mount paper on cardboard and fold it around the edges properly. We assume a sheet of letter-size paper; the procedure is the same as you would use to make a book cover.

Edge preparation

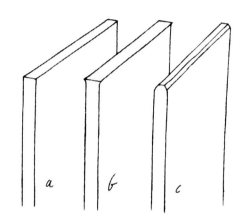

Successful mounting depends on the preparation of the edges of the cardboard. The illustration shows: (a) The desired appearance. (b) If the edges are raised from the cutting process, they should be smoothed with a bone folder or with sandpaper if necessary. (c) Edges should never be rounded!

Six steps

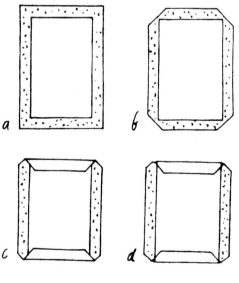

Procedure:
- a. Apply adhesive and press the paper onto the cardboard.
- b. Trim the corners.
- c. Fold over upper and lower sides.
- d. Pinch in the four corners.
- e. Fold the left and right sides.
- f. Apply paper to the back side to counteract tension. Endpapers serve this purpose in a book cover.

Width of overlap

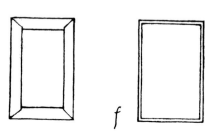

An overlap of about ½ in (12 mm) seems most pleasing. Larger formats require more overlap, but much less appears stingy and is also difficult to glue down.

The paper that you have chosen as a covering material is positioned for gluing as described on pages 47–50.

Thickness of the covering

Choice of adhesive

The thickness of the material is a factor to be considered. Heavy paper is difficult to handle, especially around the corners. It has to be wet thoroughly. For very lightweight paper or cloth use a watery adhesive mixture.

Setting the cardboard down

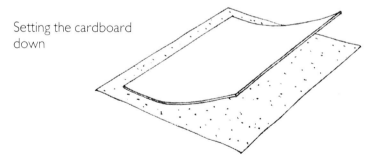

After the adhesive has been applied to the paper, set the cardboard down onto it and rub on it. Then turn the piece over, rub it again from the paper side, and turn it back to its original position.

Avoiding problems

While the cardboard is face down, the adhesive on the edges of the paper may be transferred to the protective sheet on the work surface underneath and spoil the paper, once the piece is turned over again. You can avoid this danger when working with smaller pieces by choosing a fresh spot to work on after turning.

Trimming the corners

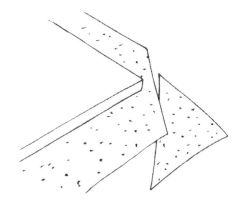

Trim the corners at an angle of 45 degrees, cutting no closer to the corner than one and a half times the thickness of the cardboard. If you cut off too much, the corner cannot be covered completely. If you leave too much, the corner becomes too bulky.

Sequence

As a rule, treat boards, portfolios, and books in this sequence: top and bottom first, sides last. In a box start with two parallel sides, followed by the other pair. For most efficient folding, position the piece at the edge of the table so that the side you want to work on sticks out over the edge about 1 in.

Two steps of folding

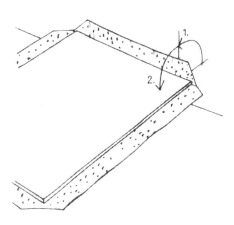

Hold the piece in place with your left hand and fold the paper up to form a right angle with the cardboard. Ideally, the paper should stick to the cut edge of the cardboard.

Caution

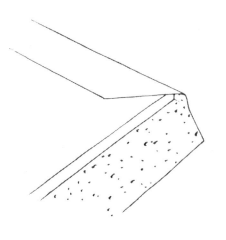

During the second step of the process rub the paper down with your thumb inch by inch, starting at the left corner. The paper should stick immediately, and the edge should look clean and even. To prevent the edge of the table from leaving an impression in the still-damp cover, position your left hand well behind the edge to avoid pressure along the edge. Another precaution is to slightly round off the edge of the table.

Hollow spots

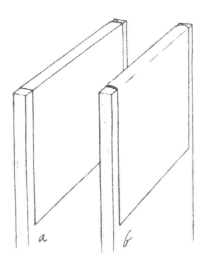

The drawing shows a right (*a*) and wrong (*b*) fold. Hollow spots should also be avoided.

Corners

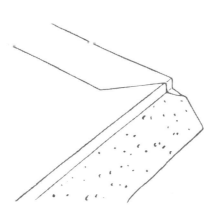

After two parallel sides are done, and before you start the remaining sides, the corners receive special treatment. Tools are fingertips and fingernails. Now position the piece flat on the work surface.

Finish

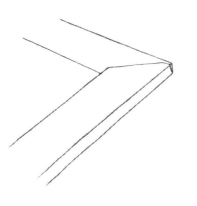

Before you fold the two remaining sides, make sure that the adhesive is still damp enough, and reapply if necessary. Proceed as with the first two sides. Check the whole piece carefully for loose spots and too-sharp corners, either of which you pat down gently with a bone folder.

Trimming the overlap

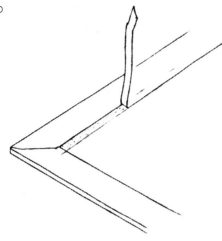

To adjust for variations in width of the folds, trim them back where necessary. Measure carefully, mark the line with a pencil, and cut, but just deep enough to sever the paper. Finally, pull the strip off at an outward angle. It is best to wait one or two minutes after folding before cutting.

Avoiding ridges

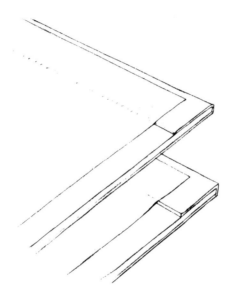

Below a backing, the untreated edge of the fold stays visible. It is best to rub down the edge to create a smooth transition, especially when thick paper is used.

Backing

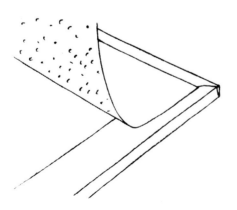

Backings, as mentioned, should be cut in the same grain direction as the front paper. Since their tensions should be slightly higher, the percentage of paste in the adhesive has to be higher too. Leave about 1 in (2–3 cm) free around all four sides.

MOUNTING POSTERS

Mounting on cardboard

Here we are talking about larger formats of about 30 to 40 in (70 × 100 cm). Even if you use extremely thick cardboard, the risk of warping and bending can never be avoided. Boards of hard wood or compressed wood are a far better choice for posters.

The following questions have to be considered:

If the poster cannot be trimmed

a. Can you trim the poster? If the answer is no, choose a board larger than the paper, particularly in the dimension that is in the direction of the grain, which has to be determined beforehand. (If scraps of the same paper type are available, use them for a test.) After the poster is mounted and dried, the board can be cut back to size by a carpenter and the edges sanded.

If the size of the board is given

b. If the board determines the size of the finished work, the overhanging edges of the paper are trimmed off after mounting. The edge of the board guides the knife. In another method the edge strips are removed by rubbing along the edge of the board with a sanding file at a 45-degree angle until the strips come loose.

Folding the edges back

c. Should the edges of the poster be folded around? For this a wider margin is needed, approximately 1 in (3 cm) plus the width of the board.

Rounding the edges of a board

It is a good idea to soften the sharp edges of the board with sandpaper, but remove dust and particles carefully before continuing the work.

Corner cuts

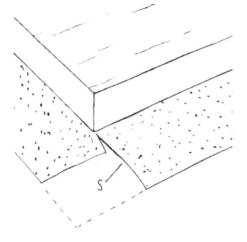

The pattern of a corner cut is shown in the drawing. The cut, s, is not a continuation of the board-edge, but it is placed slightly inward and parallel to the edge.

Position of corner cuts

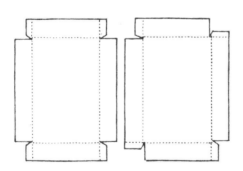

The left-hand drawing shows the correct arrangements of corner cuts; the right-hand drawing illustrates an incorrect arrangement.

Folding

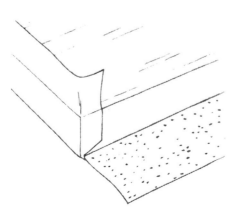

First the side with the slight overlap is folded, but only onto the edge of the board.

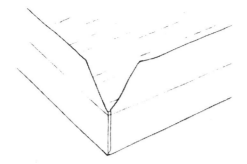

When the other half of the corner is folded up, paste the rest of the edges to the back of the board, after making V-shaped cuts as shown in the illustration. The cuts do not meet exactly at the corner of the board but slightly above it.

Covering the sides of the board

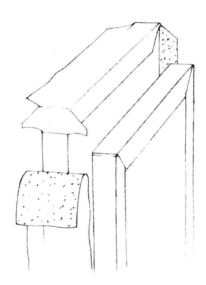

d. Should the edges of the board be covered with separate strips? Paper or cloth strips have to be mounted before the poster. The drawing shows the process in compressed form. The corners should be folded as shown on page 59.

Boards with smooth surfaces should be treated with sandpaper or with a grease-dissolving detergent, especially if the adhesive does not contain PVA.

If the poster and board are the same size, the board can also be used as a work surface for gluing.

Choice of adhesive

A 4:1 mixture of paste and PVA is advisable for average papers. Pure paste carries the least risk. Conspicuous spots can be removed, and the paper can be repositioned if necessary or even lifted off the board for some time after it has been rubbed down.

Applying the adhesive

The adhesive can be applied with a big round brush, or a roller made of foam rubber or lamb's wool. Always allow the paper to expand before positioning it on the board.

Positioning	The large formats we are discussing here can only be positioned properly with the help of another person. They have to be set down along an edge parallel to the grain direction. Hold up the rest of the sheet and lower it down slowly, while rubbing it down. If necessary, protect the surface with an additional layer of paper.
Rubbing down	
Lining	A lining on the back to counteract the pull of the poster is always necessary, even when ¾ in (2 cm) thick board is used. As usual, the lining has to have the same grain direction as the front sheet.
Weighting	If you mount the paper on cardboard, use weights during the entire drying process.

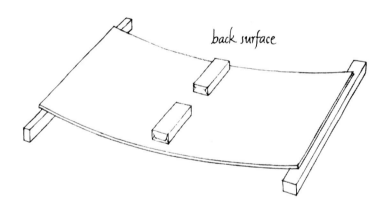

Drying	Cardboard with one smooth surface has a way of warping so that the smooth surface becomes concave. A remedy is shown here. Raise two edges on pieces of wood and place weights in the middle.
Drying in open air	

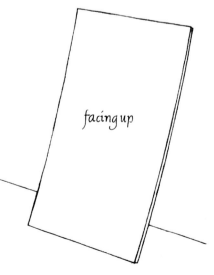

Wooden boards can be dried by leaning them against a wall with the front facing the wall. Whether a piece stays flat or not also depends on the

Warping

climate. Putting a piece over a radiator or exposing it to strong sunlight will almost certainly cause warping.

Stabilizing

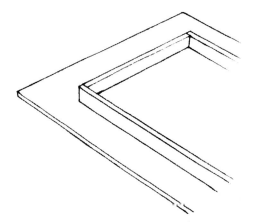

A wooden board can be stabilized considerably by the addition of a frame structure on the back. This can also provide a place to mount hangers. Even two parallel strips of wood mounted on the back can have a similar effect.

MOUNTING PAPER ON PAPER

Paper is often glued to paper. For example, a thin or damaged print may need to be reinforced with another piece of paper, or two sheets of different color are to be combined into a single sheet.

Gluing

Gluing two sheets of paper together is no easy task — if the product is to be flat. Take two medium-weight sheets of paper of similar quality, position them side by side, and apply paste to both equally with a round brush and again after 3 minutes. A waiting period of another 3 minutes gives the paper the crucial opportunity to expand evenly — a prerequisite for success.

Setting on

Set the paper down as described on page 50. No force can be used, and the rubbing has to be done through an additional layer of paper, since any side shifts and tensions will show up as waves after drying.

Rubbing down using a protective sheet

In the same way as you put to-

Making cardboard from paper

gether a sandpaper file (see below), you can actually make cardboard by combining any number of sheets of paper. It is of great importance not to cut short the drying process. A day seems the minimum (see page 53). An addition of PVA to the paste will help yield good results.

Another sequence is to combine several sets of two glued-together papers, add these to each other, and so on.

Cardboard was not available commercially until the middle of the eighteenth century. Bookbinders made their own and they often used older manuscripts as raw materials. We owe the survival of many a precious document to this practice, which preserved what might otherwise have been destroyed.

MOUNTING CARDBOARD ON CARDBOARD

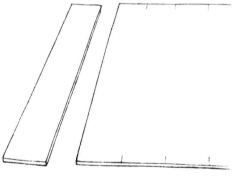

A sandpaper file

Cutting

During this exercise, we produce a useful tool, a sandpaper file. Cut pieces of 2 × 10 in (5 × 25 cm) from thin cardboard.

Gluing

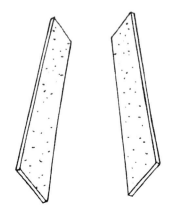

Brush PVA on the rough sides of two of them and glue them together. There should be no traces of adhesive visible at the sides.

Assembly

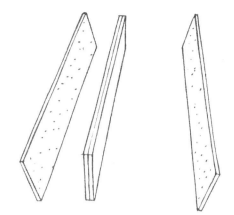

Place weights on the piece and put it aside. Then glue a third and then a fourth piece on each side. Again weight the whole piece (see page 53).

Drying

Like the gluing process, the drying process has to be symmetrical, which means the moisture should be able to escape evenly on both sides. If you use PVA there is no rush, for the pieces will still adhere to each other after 10 minutes, but more pressure may be needed, since they could already be slightly warped after this time.

If you wonder why this job is done with unthinned PVA, try the same project with a 1:1 mixture of paste and PVA. The result will clearly demonstrate the reason.

Evening the edges

After about ½ hour the edges can be evened out on a piece of sandpaper (see page 26). Then cut two fresh strips of sandpaper of slightly smaller dimensions than the cardboard. The grain of the strips should be different, one fine, one coarse.

Cutting sandpaper

Sandpaper cannot be cut like regular paper. Score on the back side and bend in both directions, then tear it in two.

Gluing sandpaper

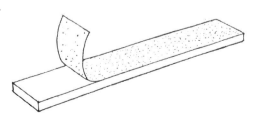

Glue the sandpaper strips with PVA, press them onto the cardboard, and leave the piece to dry under pressure. After about an hour the sandpaper file is ready.

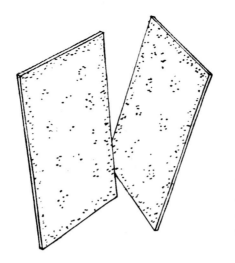

Gluing large pieces of cardboard

If large pieces of cardboard are to be connected, apply the adhesive along a 1 in (3 cm) wide rim around all sides and on spots scattered across the middle area. Higher stability and less tension will result.

STRETCHING PAPER

Cutting a pattern

In some cases it is desirable to mount only the edges of a sheet of paper, not its entire area, in order to stretch it. To prepare a sheet of any given dimension cut a piece of paper that measures ¼ in (5 mm) less in length and width than the sheet you plan to stretch. This template, placed in the middle of the sheet, allows you to apply the adhesive along exact lines.

Choosing adhesive

Stretching paper over cardboard

Use unthinned PVA and work swiftly. If you want to fold the edges of a stretched sheet of paper onto the back of the cardboard, apply the adhesive to the cardboard as before. To center the applied paper it is useful to mark the dimensions of the cardboard onto the back of the paper at the beginning of the process. Cut off the overhanging corners at a 45-degree angle. Each flap should be treated separately; see drawing on page 259. Corners require special care, as before.

In this method the flaps should also be glued only along the inside edges.

Stretching paper for painting

If you plan to paint with tempera or watercolor on a piece of paper, it should first be stretched on a board. Place it on the right spot, hold it in position, and glue it down with four strips of paper that correspond to the sides. After 15 minutes dampen the sheet evenly with a sponge, but avoid the glued strips. When the paper dries, it will contract and become smooth. If edge strips cannot be used to attach the paper to the board, the stretching method described above can be used.

GLUING TISSUE PAPER

Extreme situations often provide insights into general principles. Without doubt, gluing tissue paper constitutes an extreme situation.

Grain direction

Determine the grain direction by ripping (the smoother edges will indicate grain direction) or by one of the other tests described on page 14. Using very thin paste and a soft brush, apply an even layer of adhesive. Transfer the paper onto a fresh sheet, and repeat the paste application.

Choosing adhesive

Applying adhesive

Setting on

Rubbing down

Using a protective sheet

This time do not set the paper down on the cardboard. The paper stays on a clean surface, while the cardboard is gently rolled onto it, starting either from an edge or the middle line. Apply gentle pressure and flip the piece to rub the paper down middle to edge, always through an extra sheet of paper. Do not apply the lining to the back immediately; wait at least half an hour and follow the instructions on page 53. Another option is to finish the back before the front.

INDIRECT GLUING

In some cases you will need to avoid direct application of adhesive to the material to be glued:
a. The paper is too small (for example, labels);
b. The paper is a narrow strip;
c. The paper rolls up;
d. The material to be glued is loosely woven fabric that would let the adhesive bleed through.

Choose the right adhesive mixture (see page 46) and apply it with a brush to a surface of linoleum, glass, or metal, or a vinyl-covered board. Place the paper or material to be glued on the layer of adhesive, taking care not to create air bubbles or creases. On top of it put a larger sheet of paper through which you rub on the first sheet.

After 1 or 2 minutes the material should have expanded and can be removed. Check for a spotless application of adhesive and repeat the process if necessary.

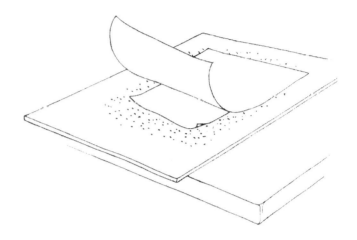

GLUING BOOK CLOTH

Use paste and PVA in a 1:1 mixture. If the ratio of paste is higher, the cloth loses its body, and if a layer of lightweight paper is attached to it, it may come off.

During more demanding projects time often is at a premium, and you barely manage to cut and fold before the adhesive dries. Here a higher percentage of paste is desirable, and compromises are necessary.

GLUING UNSIZED CLOTH

Iron the cloth until smooth. If the weave is dense enough the adhesive can be applied directly, but the method described on page 70 is safer. Use a mixture of PVA and paste in a 4:1 ratio and apply it to a smooth surface; pull the cloth across it until it is covered but not saturated by the adhesive.

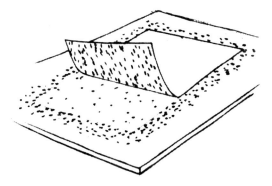

If you want to use unsized cloth and size it yourself, use paste with a small addition of PVA and apply it until the cloth appears completely covered on one side. Dry thoroughly before proceeding. Since the fabric will contract, make sure to allow for shrinkage beforehand.

Paste is the adhesive of choice since spots that bleed through usually disappear after drying, whereas PVA stays visible.

GLUING A SERIES OF SHEETS

Layering

If a whole series of sheets of paper is to be glued at the same time as, for instance, in making your paper collection, it is best to treat them as a unit. Layer about half a dozen sheets spread about 1/8 in (2 mm) apart; resist the temptation to glue a wider area.

Covering

Adhesive

Grain direction

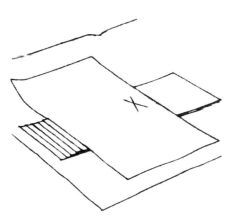

To ensure equal treatment of the topmost sheet use an extra sheet and hold it in place with two fingers at the spot marked "x" in the drawing. Choice of adhesive depends on the kind of paper used; see the chart on page 46. The sheets will warp less if they are arranged in the same grain direction. If the sheets are illustrations for a bound book they can be glued along the upper edge or the inner side of the page.

Weighting

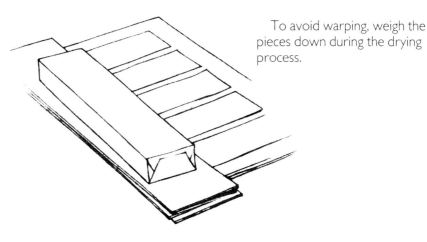

To avoid warping, weigh the pieces down during the drying process.

THE PLATONIC SOLIDS

"Our eyes are made to see the shapes under the sun: Light and shadow reveal forms; cubes, spheres, cylinders, or pyramids are the primary forms that the light makes visible. These images appear clear and tangible, and therefore they are not just beautiful, they are the most beautiful things we know. On this all agree, the child, the savage, and the philosopher."

Le Corbusier. *A View of Architecture*, 1922.

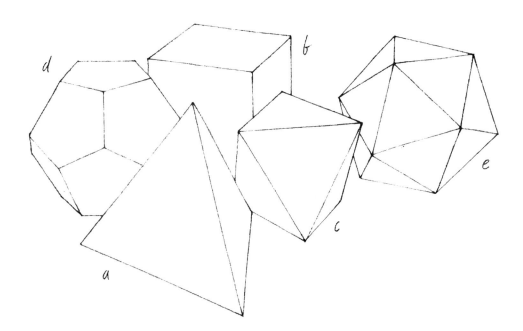

a. The tetrahedron, or regular four-sided solid.
b. The hexahedron, or regular six-sided solid.
c. The octahedron, or regular eight-sided solid.
d. The dodecahedron, or regular twelve-sided solid.
e. The icosahedron, or regular twenty-sided solid.

GENERAL AND SPECIFIC PROPERTIES

Next we construct the so-called regular solids, not just to improve manual skills, but because these structures are highly interesting. There are five of these Platonic bodies, as the regular solids are also called, no more, no less. On each of them we find equal and regular polygons with matching sides and angles. The geometric shapes that compose their sides are triangles, squares, and pentagons. In spite of these similarities each of the solids has its particular characteristics, which become obvious upon examination. Close observation and a comparison of their outward appearances should prove quite fascinating.

Take a cube, for example, a clumsy and familiar thing, its surprises hidden in the interior, and compare it to a four-sided structure: pointy, stable, tall, and proud — a pyramid. One should write a family history of these fine fellows.

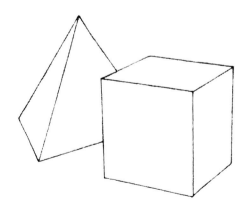

Solids and nets

To construct a multisurfaced form we start with its net, which is nothing but a two-dimensional representation of the entire surface. We determine the shapes that make up the surface, draw them and the necessary folds for gluing onto the paper that is to be used for the construction, cut, fold, and finally glue the piece together.

The result is a transformation from a two- to a three-dimensional object, from a flexible sheet of paper to a stable structure. If this process is fully experienced, the following steps will be all the more enjoyable.

Similarities and contrasts

Even though the regular solids may appear very different at first glance, there are definite relationships among them. In each one of them we can find the governing structures of the others. Take the case of a cube and a tetrahedron, in which there is not a single right angle. If the edge of the tetrahedron is the same length as the diagonal of the cube's side, the tetrahedron will fit exactly into the cube. The four empty spaces left will have the form of three-sided pyramids.

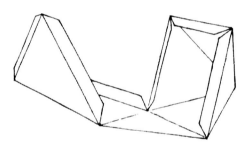

Choosing paper

Because of the already mentioned stabilizing effect of the folding process, relatively lightweight materials will be sufficient for our purposes.

Concentration is a prerequisite. Inexact work produces structures that cannot be completed, edges that will not fit, and other similar and obvious problems.

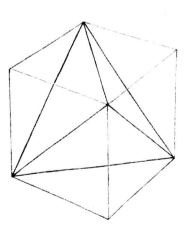

HISTORICAL FACTS

The shapes we are talking about are not man-made. They are as ancient as creation itself, and they reveal to us universal laws of nature. Present everywhere, they appear in their most obvious forms in crystals.

The regular solids in nature

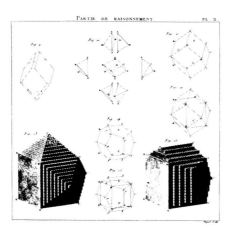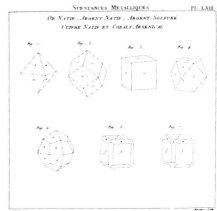

C. Haüy

This illustration from C. Haüy's book, *Traité de Minéralogie*, (Paris, 1801), shows mineral structures.

Minerals and gems have most certainly attracted man's interest from earliest times. We do not know what importance they were given then, but many centuries ago in Greece the Pythagoreans venerated crystals as representations of the gods. Their efforts to investigate them produced the foundation of mathematics in the western hemisphere.

Pythagoras

The Renaissance

When the world of the Greeks was rediscovered during the Renaissance, men such as Leonardo da Vinci and Albrecht Dürer were fascinated by these simple but mysterious objects. Numerous tracts and sketches bear witness to this fascination.

Leonardo da Vinci

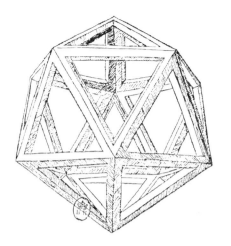

Leonardo da Vinci drew this icosahedron as an illustration to a book by Luca Pacioli, the Italian mathematician.

Albrecht Dürer

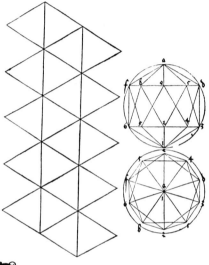

These renderings of an icosahedron come from Dürer's *Unterweisung de Mesung* (Measuring Instructions) and show his fascination with proportions.

The regular solids in astronomy

Johannes Kepler

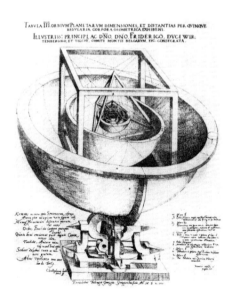

In 1596 the twenty-four-year-old astronomer Johannes Kepler had the ingenious idea of calculating the distances of celestial bodies with the help of the Platonic bodies.

This model shows Kepler's arrangement of the solids. Starting from the outermost shell we find a cube, a tetrahedron, an octahedron, and an icosahedron in a set sequence. They are shown with their inscribed and circumscribed spheres; for obvious reasons, only half is visible. Each of the geometric figures is surrounded, or circumscribed, by a sphere that touches all its corners. A second sphere touches the middle of all sides of the bodies; this is called an inscribed sphere.

Inscribed and
circumscribed spheres

With the exception of the outermost sphere, each one serves as inscribed and circumscribed sphere for two different bodies at the same time. All of them have the same center, which represents the sun in the model. The distance from the center to the various spheres was assumed to represent the position of the orbits of the planets.

Alexander Graham Bell

Towards the end of the last century, Alexander Graham Bell, the inventor of the telephone, successfully constructed kites and flying machines utilizing theories derived from polyhedra.

Buckminster Fuller

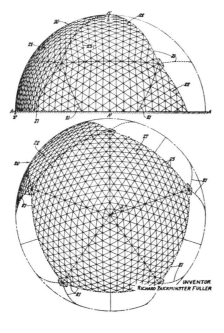

In this century, large dome constructions of metal and glass have been built by designers such as Buckminster Fuller.

All of these applications utilize different aspects of the laws that can be derived from the regular solids.

EDUCATIONAL USES OF PAPER MODELS

Art education

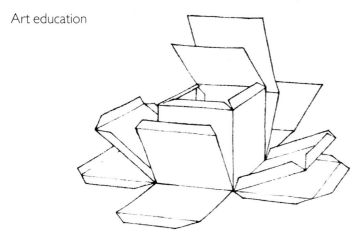

Perspective

Mathematics

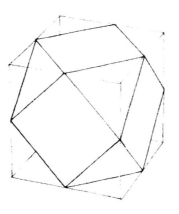

To make or, rather, to construct a paper model of a geometric solid is a very inspiring process. The evolution of a two- into a three-dimensional structure will be of interest to an inquiring mind. (My first book, about paper, offers illustrations and descriptions of various playful approaches to cube constructions.) The other four solids lend themselves to similar approaches. The way from tetrahedron to icosahedron leads to the sphere.

The volume, in relation to the surface, increases steadily. The cube is the body that combines a maximum of surface with the smallest volume; the sphere yields a maximum of volume with the smallest surface.

Constructions of these solids requires skill and care. They help explain perspective, light, and shadow, and the relationships between different shapes.

Paper models of geometric bodies can also be of great help to the mathematics teacher as a visualization of abstract problems.

Importance of
the senses

Spatial sense

We are in the process of rediscovering the importance of our senses today. Sight and touch, as well as movement, open the surrounding world for us and let us experience space. Our ability to relate to others depends on them. It is an increasing necessity to dam the flood of two-dimensional impressions delivered by the mass media. To do this we have to learn how to make, build, and shape things, how to draw, to recognize forms, colors, and motions. Basic human skills need to be resurrected. The primal shapes of our regular solids are the most elementary forms imaginable.

BUILDING A PAPER MODEL

Preparations

Size relationships

Before you can start to build a model, you must determine the size and relationships of the forms. Unless there is some special consideration, make your models about the size of a large apple. Very large or small models are cumbersome, especially for beginners.

To relate various bodies to each other note and compare the following criteria:
   a. length of the sides
   b. surface area
   c. volume, mathematically derived
   d. volume, estimated

To match estimated volumes it may be necessary to construct several pieces until you obtain an optically pleasing result. The following measurements will produce visually equivalent volumes: sides of 6 in (15 cm) for tetrahedrons, 3¾ in (9.5 cm) for cubes, 4¼ in (11 cm) for octahedrons, 2 in (4.7 cm) for dodecahedrons, and 3 in (7.5 cm) for icosahedrons. Volume can appear different from different views.

The procedure described for one can be used for all of the solids. Most of the following drawings require a basic knowledge of geometry. A little review of high-school skills will do, but exactness cannot be compromised, because small errors eventually add up to big ones.

The construction itself is usually a pleasurable process, provided you are able to avoid focusing on the repetitiveness and enjoy a new aspect of the familiar.

Construction

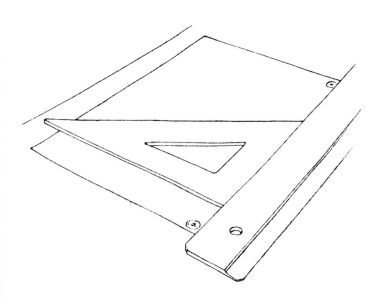

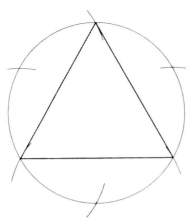

Constructing a triangle in a circle

Constructing a triangle with a given side

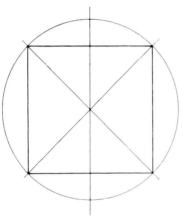

Constructing a square in a circle

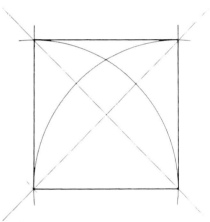

Constructing a square with a given side

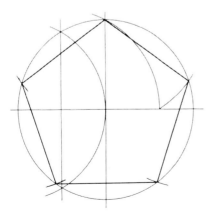

Constructing a pentagon in a circle

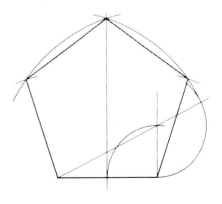

Constructing a pentagon with a given side

THE TETRAHEDRON

Properties

The tetrahedron is composed of four equilateral triangles and rests on one of its sides. It looks like a pyramid. Inscribed in a cube, each of its edges is a diagonal in one of the sides of the cube. There are no diagonals in any dimensions in the tetrahedron, and it is the only polygon with the same number of corners and sides. Check! Three squares of equal size cut the tetrahedron in half and cross each other. They form a "skeleton" for the inscribed octahedron.

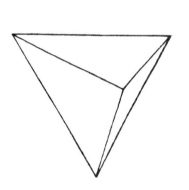

A net should not be constructed in a more difficult way than necessary, but on occasion one or more small parts of a piece are easier to handle than the whole.

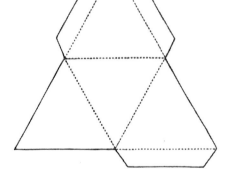

Impractical net

A net can look familiar but contain hidden problems. Here three free edges meet in one corner, which would necessarily present difficulties.

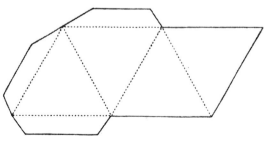

Practical net

Even though this net seems less obvious, it is preferable by far.

Position of folds

Model

To determine the optimal position for the folds, a model with a side length of about 1 ½ in (4 cm) can be of great help. Mark the corresponding edges and flatten the model. Keep the edge that will be glued last free of a fold, which will be positioned on the opposite edge (an explanation follows). Such a model is also useful to determine if the arrangement of the sides should be changed.

Simple fold

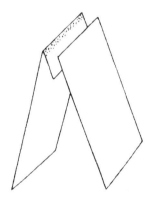

The drawing shows how two pieces of paper can be joined by a fold.

Material

For a polygon of about 4 in (10 cm) in diameter, medium-weight paper seems best for creasing and heavy paper for scoring. White paper presents the shapes most clearly, but colors should by no means be excluded. It is best to choose the lightest-weight paper that will do the job. It makes folding easier and more precise, and the stabilizing effect of the folds becomes more obvious

Estimating the size

Drawing the pattern

Creasing

Pre-folding

Using your model as a guide, estimate the size of the paper needed and cut the necessary rectangle from the sheet. All tools should be in working order, all pencils sharpened. Make your marks not with random black dots, but with exact lines about 1/16 in (1–2 mm) long. First crease where necessary, then cut. If the order is reversed, you cannot achieve the correct intersection of lines.

There is a certain temptation to omit this step, but pre-folding saves considerable trouble during gluing and it prevents bulging sides in the finished piece. Fold all sides and along the attached edges, and rub them down gently with a bone folder.

Trial folding

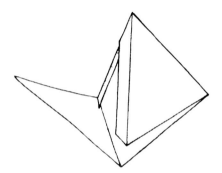

Now fold the piece, to make sure that all the folds are in the right place, and that they are trimmed at the sides to avoid double thickness.

Sequence of gluing

Applying adhesive

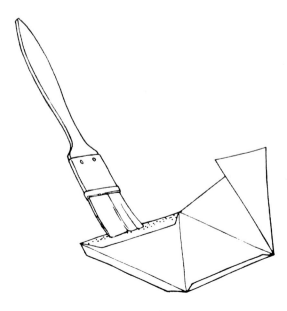

When you have corrected any mistakes, the gluing can begin. This is a process that requires great care. It is helpful to mark the sequence of steps on the model, since concentration on the gluing itself may easily divert you from the chosen sequence.

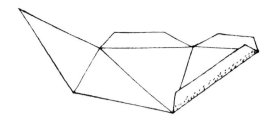

With a 1 in (3 cm) flat brush, apply unthinned PVA to a ⅛ in (3 mm) strip of the first fold. A shaggy brush will make this job difficult. Experience will show why a ½ in (10 mm) fold is needed for a ⅛ in (3 mm) gluing area. If adhesive is applied to the entire width of the strip it will warp, and a fold of ⅛ in (3 mm) would be impossible to handle. After a 30-second drying period, combine the two parts and hold them together for about a minute. Wherever possible, pressure should be applied from the outside as well as the inside of the piece. If the adhesive forms droplets or if the paper warps, you have used too much PVA.

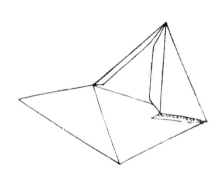

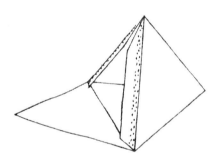

When it seems that the edges will not shift or, worse, separate any more, proceed to the next edge. When you get to the last side, it will be necessary to close two edges at the same time. Since the inside is now inaccessible, bend the folds up carefully to provide some resistance during the gluing process.

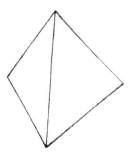

The glued piece should appear as crisp and clean as it was when you first folded it. Errors such as cuts, partially open glue joints, traces of adhesive on the outside, or dirty smudges are almost impossible to correct. The only remedy is a fresh start.

The color illustrations on page 129 show some variations on the basic model of the tetrahedron.

THE CUBE

The cube is composed of six squares. All its angles measure 90 degrees; it has twelve edges and eight corners. It is the most familiar of regular solids.

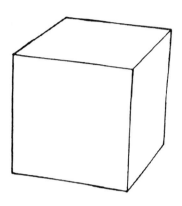

The small black and white die with rounded corners and edges that rolls across game boards connotes quick action, while a large cube represents weight and immobility. A second look reveals that interesting features are hidden by the simple appearance of the cube.

Some examples of cubes are shown in the color illustrations on pages 130–131.

If different nets are possible, to find them all is a game in itself.

Use the familiar cross-shaped net shown here. Crease, and provide folds for attaching the edges.

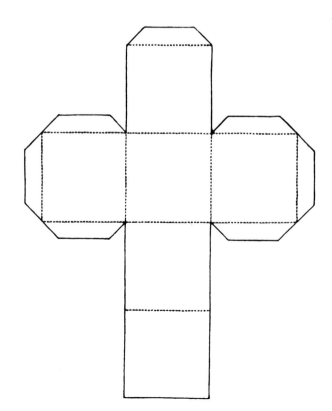

THE OCTAHEDRON

The octahedron is formed by eight equilateral triangles. It has twelve edges and six corners. The angles of its net all measure 60 degrees.

The octahedron exemplifies how an object can look different if the point of view changes. Suspended from one corner, it looks like a double pyramid, and its middle plane seems emphasized. When it is lying on a table, however, the two adjacent triangles capture your attention, and the other six sides are subordinated. Similar observations can be made about other solids.

To construct an octahedron we use a new technique. All edges are supplied with a fold, so that you connect each pair of sides with a double fold. The advantages are as follows:
1. The moisture in the adhesive cannot be transferred to the outside of the polygon.
2. The position of the finished overlaps provides greater stability.
3. Perfect joints are possible, even where they cannot be rubbed down.
4. No shadows or folds will appear on the finished piece.

Double fold

THE DODECAHEDRON

The dodecahedron is composed of twelve pentagons. It has twenty corners and thirty edges.

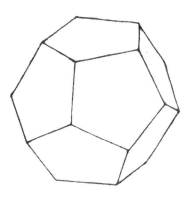

Using double folds, we construct two identical halves. Crease, which increases precision but requires sturdier material. For a dodecahedron with a diameter of about 5 in (12 cm), use a lightweight (200 gm^2) cardboard. Prefold the two halves and combine them one fold at a time.

Start with two nets of six pentagons each. Construct the net from the outside inward, not from the center pentagon out; this yields a better result.

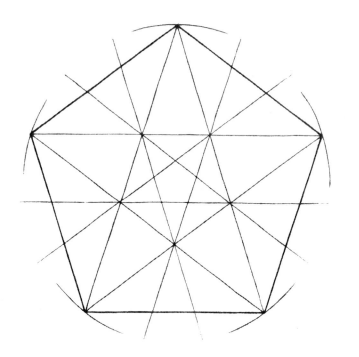

THE ICOSAHEDRON

The icosahedron is composed of twenty regular triangles. It has twelve corners and thirty edges.

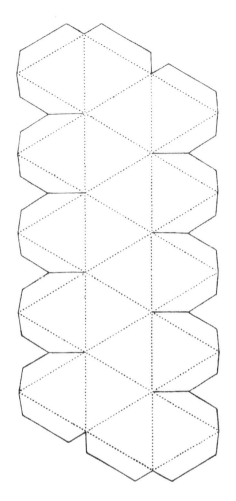

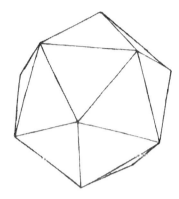

Score the icosahedron and use double folds. The net of the icosahedron, quite unlike that of the dodecahedron, resembles a saw.

The two outside rows of five pentagons each are assembled first; stability comes with gluing the remaining edges.

LARGE PAPER MODELS

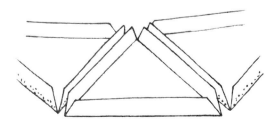

The advantages of double folds become most obvious with large polyhedra. For an icosahedron with sides 20 in (50 cm) long, prepare twenty equilateral triangles with folds and put them together. If you work carefully, you should be able to surmount all the difficulties, even with large pieces.

THE ARCHIMEDEAN SOLIDS

The way leads from the tetrahedron to the icosahedron to the sphere, just as in plan a dodecagon is closer to a circle than a square.

Moving beyond the icosahedron, we find no sixth regular polyhedron, but "cutting off" the twelve corners of an icosahedron we come closer to the shape of a sphere. We now have a solid formed by regular pentagons and twenty hexagons. If the "cuts" are placed accordingly, the hexagons will be regular. The new solid has thirty-two sides and two different kinds of regular polygons. Every regular solid can be treated this way. If the resulting sides are regular polygons, they will form a semiregular polyhedron, which is sometimes called an Archimedean body.

It is easy to imagine that at long last we would end up with an approximation of a sphere if we continued to "cut off" corners. It is, however, time-consuming enough to produce the model of a thirty-two-sided solid, and anyone but the most ardent mathematician would lose interest. I value perseverance, but I counsel against projects of gigantic dimensions, since routinely executed manual labor is not what we are after. Besides, it is often the solid with the fewest sides that provides greatest visual interest and inspires experiments.

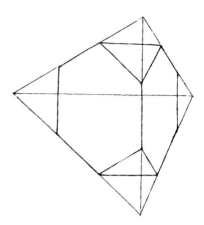

Combination of two tetrahedrons

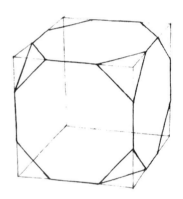

Combination of cube and octahedron

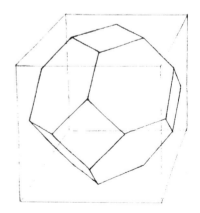

Combination of cube and octahedron. The nets of these solids can be constructed following the instructions on page 84.

INSIDE A CUBE

The drawings below illustrate my earlier claim that there are surprises hidden inside a simple cube. What happens to a tetrahedron or an octahedron in a way also happens to the cube into which they are inscribed. Simply to draw these solids should provide valuable insights. In time their separation and combination can become a mental exercise.

These freehand drawings show the process of splitting the existing solids or of recombining the resulting pieces.

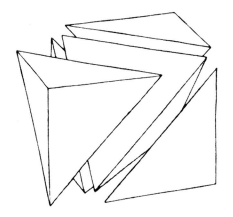

Taking off the corners of a cube reveals a new inner tetrahedron, a regular polyhedron.

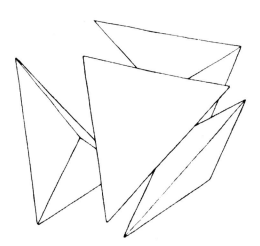

Separating the tetrahedron into four equal parts produces four three-sided pyramids.

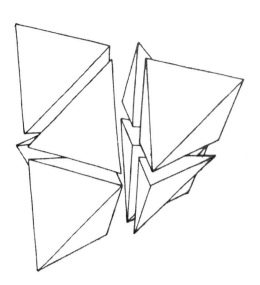

This tetrahedron is separated into eight parts by three cuts.

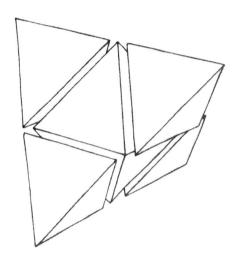

Four cuts can remove four regular tetrahedrons from the original one and leave a regular octahedron in the middle. The edges of this resulting solid match the "edges" of the three planes along which the original tetrahedron was halved.

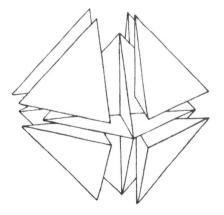

Three cuts in different planes will separate this octahedron into eight equal, regular, and three-sided pyramids. The separation scheme is included in the drawing because it is a pleasing structure in itself: three intersecting squares, representing the three separating planes of a tetrahedron.

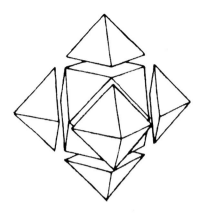

The eight corners of an octahedron can be cut off to leave a cube-octahedron. It is easy to see how a cube could be formed from it.

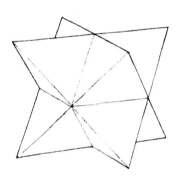

The so-called Kepler star results from two intersecting tetrahedrons of equal size. The common space has the form of an octahedron; its twelve edges are the receding edges of a star.

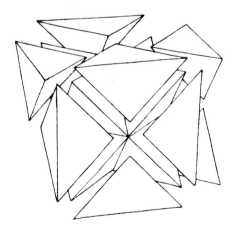

We can also imagine that a Kepler star could come from a cube. We would have to cut off twelve irregular solids.

THE MAT

Mats, usually made from lightweight cardboard, serve as frames for prints, watercolors, drawings, and photographs. The purpose of a mat is not to upgrade a mediocre work, but to show a piece to its best advantage. Simple examples would be: An art teacher prepares for a job interview and presents samples of his work in mats and portfolios (see page 175); someone has saved her grandfather's travel sketches and makes a hinged box for them (see page 158); a collector of graphic art wants to display a changing selection of pieces in frames but store the rest in a box with a flap (see page 124). The solutions in all of these situations can be both beautiful and useful.

A mat should never distract from what it is meant to enhance. Therefore, strong contrasts should be avoided. It is obvious that a colored drawing will be influenced by the color of its surroundings, and even a good piece can be all but destroyed by the wrong mat.

The rules are few and flexible; each situation should be approached as a new one. The case of the art teacher is a simple one. He is probably best advised to choose an off-white bristol board for most of his pieces and to make a few mats with carefully matched colors for special ones.

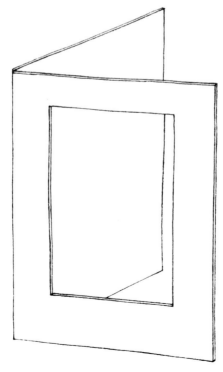

1. The simplest kind of mat is a folded piece of cardboard with a cutout window.
2. Another type of mat consists of two separate pieces, joined by a hinge of paper or fabric. This makes it possible to use different kinds of material, for instance a thinner and cheaper one for the back than for the front.
3. The third version resembles the second, but it features a covered front, to match it to the displayed picture.
4. Multiple layers of front mats should be left to the professional.

PROPORTIONS

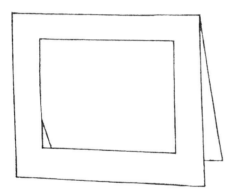

There are no hard and fast rules governing the placement of the cutout. The work to be matted is usually optically centered, and it should not be placed below this position because it would appear to slide down. The drawing shows the window slightly above the optical center of the mat.

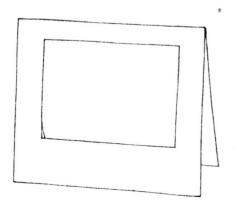

Some people prefer a greater contrast between upper and lower rim, but caution is advised. The drawing shows a shift that has gone too far.

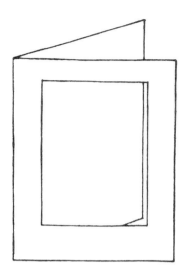

Others approach the question mathematically and choose proportions of 3:4:5 (for upper, side, and bottom margins). This seems to yield more pleasing results than a ratio of 4:4:5, which produces equal widths for the top and the sides.

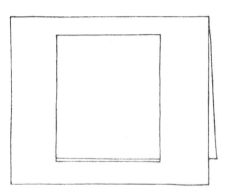

A vertical piece can look good in a horizontal mat. On the other hand, the opposite arrangement may be chosen with equally good effect.

THE FOLDED MAT

Thickness of material

Choice of material

Reaction to light

Acid-free board

Cutting to size

Window dimensions

    The maximum thickness of the material for a folded mat is that which will still allow clean folds.
    Bristol, cover, illustration, and mat board are all possible choices for mats. You may chose cardboard that is smooth or matt, bright white or off-white, or colored. Choose the one best suited to your purpose.
    If the mat will be exposed to daylight, a nonyellowing material should be chosen. This means that the material will have to be acid-free, and probably more expensive.
    You can cut the mat to the right dimensions before or after folding. The first case requires exact folding, the second a sheet about 1/8 in (2 mm) larger on each side than eventually needed (see text on folding and cutting).
    It is a common error to cut the window too small. This diminishes the effect of the artwork. Do not cover up an inch too much, even if it means extra work.

    Look at your artwork attentively. Take four strips of paper similar to your mat material and move them around the four sides until the result pleases you. Only then should you take measurements and proceed.

Edges

Cuts should be clean and straight; no fuzzy edges should appear anywhere. Use a bone folder or a fingernail to give the cutout a finished appearance.

MOUNTING A DRAWING IN THE MAT

Hinges

Placement

Gluing

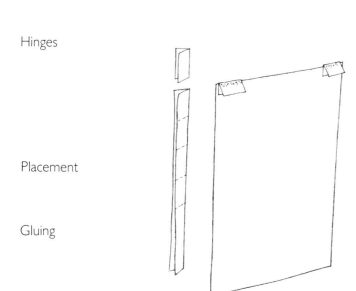

Serious damage to the artwork can result from errors during this step. A valuable piece of art must never be glued either on its entire back or along the sides, not even on all four corners. Instead, attach little hinges. You know that paper reacts to the environment, and you have to take this into account. Average-sized sheets are attached by two hinges on the upper corners, so that they can be moved to some extent.

Follow this procedure: Place the artwork under the mat and position it carefully. Any necessary changes in the window size have to be made now. When you are satisfied, weight the piece to prevent it from slipping, and apply a thin strip of adhesive to both sides of the hinges.

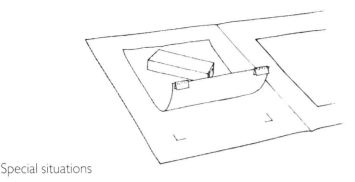

Special situations

Place the hinges between the artwork and the mat and weight them for a while. This process is very similar to the one stamp collectors use to glue stamps into an album. Larger pieces can be attached with several hinges, but never with a long and continuous one.

Sometimes the artwork to be matched cannot be trimmed any further or you want its edges to be visible. In this case make the window size just barely larger than the artwork and attach it on all corners. This presentation is recommended if the piece is to be shown in its entirety, or if a deckle edge should show.

THE UNCOVERED TWO-PIECE MAT

Choice of material

The types of cardboard mentioned earlier are choices for the front of the mat. The back can be cut from gray cardboard or other wood-containing boards. However, valuable artwork should only be presented between acid-free boards to avoid yellowing, deterioration of the paper, and foxing.

Valuable artwork

Cutting the cardboard

Cutting the window

Whether the front and the back parts of the mat should be of different or the same weight depends on the circumstances. If you use two pieces, they have to be cut separately, and the window opening is cut before the halves are assembled.

Connecting the front and back

Align the two pieces of board on a flat surface using a metal straightedge as a guide. If one of the pieces is thinner than the other, raise it to equal height with an underlying piece of board. Weight both pieces to fix their positions.

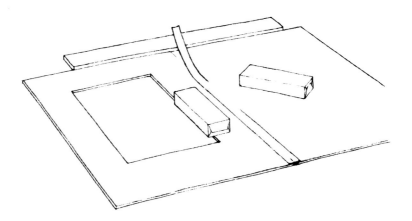

When you are working with lightweight materials, you can use paper for the hinge. For heavier boards, woven material is preferable.

Whatever material you use (except for handmade paper), you must consider the grain direction. It should be parallel to the edge of the mat.

It is a good idea to mark the spot where the hinge will be positioned, since it should end up straight and aligned with the edge.

For information about gluing, see pages 46–50.

Drying

Close the mat and weight it for about an hour. See page 53 about weighting.

The hinge can also be dried with the mat in an open position, but in this case both halves need to be weighted.

THE COVERED TWO-PIECE MAT

If heavy cardboard is hard to handle comfortably, choose lighter materials. Both pieces should be slightly larger than necessary, since they will be trimmed after they are covered and dried.

A variety of white and colored papers can be used, but they should always complement the artwork. Loud colors or strong textures could be too distracting.

How to cover the front (cutout) and the back mats is explained on pages 47–54; folding the mat and cutting the window on pages 99–102.

THE BEVELED CUT

To cut a beveled window, use a metal straightedge and secure it to your work surface with a clamp. Hold the knife at a 60-degree angle rather than vertically and cut on the upward-facing side. Needless to say, a steady hand and practice are necessary for a clean cut. Corrections are rarely possible because they are almost always visible.

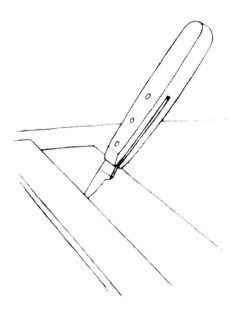

Usually a few short cuts are necessary in the corners to free the rectangle that has been cut out.

If you use thicker board, a beveled cut will expose any differences in coloration between the inside and outside layers. Do not attempt to hide this effect. Try instead to use it as a part of the design.

BOXES

Most of us have memories of boxes containing childhood treasures: buttons, snips of fabric, toys. Some of them may even have had compartments or movable flaps.

Hardly any antique boxes have come down to us. Cardboard very rarely lasts for more than a hundred years.

Nevertheless, a beautiful box can be an object of joy, even if it is less than luxuriously furnished. We use boxes for all kinds of things. They can house treasures, letters, tools, seashells, flower seeds, photographs, coins, pencils, and much more.

A box has to be practical and fitted to its purpose. The cover must open and close freely, the contents must be accessible, and the construction has to match the intended use. After all, there is a difference between keeping nails and screws in a box and storing a collection of feathers.

The paper used to cover a box does not necessarily have to be patterned. The color illustrations (see pages 133–135) show examples of simple yet very attractive boxes.

Since heavy cutting equipment is not a requirement for the work in this book, we will build only creased and scored boxes. This is by no means a disadvantage; it is even preferable.

The aesthetic possibilities are unlimited, and the boxes can be made as sturdy as necessary. While taking into consideration all possible shapes, I have come to the conclusion that round and rectangular boxes yield the most pleasing forms. Ovals and three-, six-, or eight-sided ones can be made, but the results rarely justify the

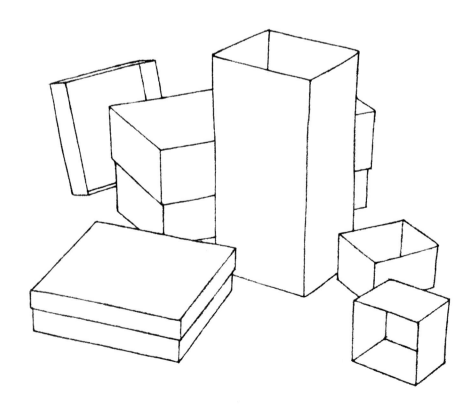

A FOLDED BOX

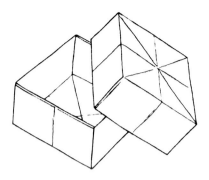

extra effort. A word of caution especially about covers that feature domes or facets: These usually spell kitsch and bad taste. A cardboard box is no chest and definitely no temple, such as the one owned by the maiden Zues, described by Gottfried Keller in his short story *Die drei gerechten Kammacher* (The Three Righteous Combmakers).

If the dimensions of a box are not dictated by its function, you can choose as a guide either mathematical proportions such as 1:2, 1:3, 3:4, 3:5, and the like, or a golden section, or standard paper or board sizes.

You may have learned how to make a folded box in nursery school. Still, the proportions are pleasing, it is interesting to build, and the little thing is even usable. In this day and age, with our preference for heavy-handed constructions, we may even benefit from some light-hearted improvisation. Graceful does not mean dainty — many graceful things are quite sturdy, as numerous examples from the plant world can show.

The little box is folded from two squares, one 12 in (30 cm) on each side, the other ½ in (1 cm) larger.

Choose medium- to heavyweight (80–160 gm^2) paper depending on the size of the intended box. The paper should be easy to fold, and the resulting box will have soft and flexible sides.

Once the squares are ready you need no ruler, pencil, scissors, or even glue.

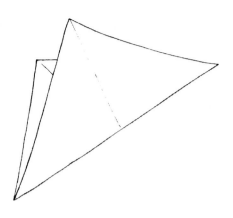

Start with the bottom of the box — the smaller square. Fold the sheet twice diagonally and open it again.

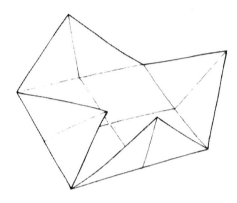

Next, fold the corners onto the middle of the square; then fold back the resulting triangles.

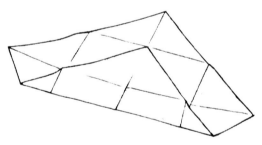

Now you have four new crossing points, and you fold each corner onto the one that is farthest away.

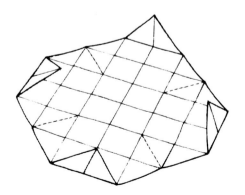

Turn over the sheet and, as the drawing shows, fold the four corners onto the nearest crossing point. Now the sheet of paper is evenly checkered. The four squares in the middle will form the bottom of the box.

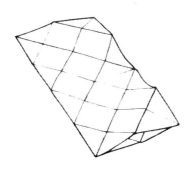

A second cross fold is needed parallel to the edges, and again on the back side of the paper. You could also fold only along the dotted lines in the previous illustration; with some practice, it is possible even to leave out all folds that will not eventually be edges of the box.

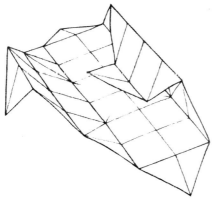

Now fold two opposing corners inward. The little upturned end triangles will rest on the bottom of the finished box. Position the walls upright.

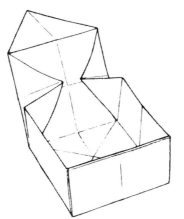

Now put the third side in place. On each side, a triangle will be folded in. Once the third little triangle is folded to the bottom, the side will secure itself to the others.

When the fourth side is done, the box is completed. Proceed in the same way with the larger piece of paper to make the cover.

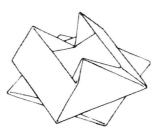

The box can be folded up for storage.

It is easy to imagine a whole set of boxes in coordinated sizes and colors.

A BOX WITH A COVER

Several variations on the basic design are possible:
a. On an otherwise unfinished box the edges can be reinforced with strips of cloth. This is a utilitarian but rather plain solution.
b. On a box with reinforced edges the sides can be covered with paper. To maintain equal tension, the inside also has to be lined.
c. Cover the outside of the box with cloth, and line it with paper.
d. Cover and line the box with paper, without edge reinforcement.

A box without any kind of reinforcement will not be usable, because the scored edges will be too weak.

The box may be planned with a cover, a flap, a divider, or various compartments.

The choice of materials for coverings and linings must not be a mere afterthought. They influence the character and even the thickness of the board by their own weight. The color illustrations on pages 133–134 show some basic ideas.

The thickness of the board is otherwise a function of the intended use and the size of the box. Generally it will vary between $1/16$ and $1/8$ in (1–2 mm), but it is possible that special circumstances would call for a large and light box or a small and sturdy one.

You should always have a mental image of the object before starting work. Each detail can determine failure or success, and only a well-made object will be a beautiful one.

Only the most general hints can be given for the choice of materials to cover boxes. It is always possible to change your plans during the course of the work. It could turn out that a certain liner matches the finished covering better than the one originally intended. Continuity and perseverance are desirable, but so are flexibility and an open mind.

The heaviest paper that can be used is one that can still be handled with precision.

It is an unfounded opinion that a solid color makes a box seem unattractive and somehow less interesting. From the past there are countless examples of works of art decorated with colorful ornaments. These days, regrettably, we see little talent for ornamentation. I advise discretion, even if it means ignoring the counsel of those arts and crafts teachers and others who urge us to be free. Even with a choice of muted colors a box can come alive, whereas a splendidly mottled one could be utterly boring. This is not to be understood as a case against lively colors, only against their thoughtless use.

The lining material has to be pliable enough to be guided into inside corners. Delicate paper is practical to use only on small boxes.

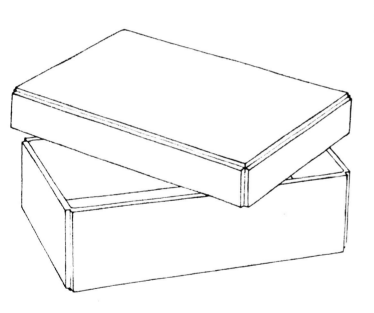

CONSTRUCTION

Flexible cardboard is preferable to brittle board. Many varieties are available either at arts and hobby supply stores or from bookbinding suppliers.

Sketch

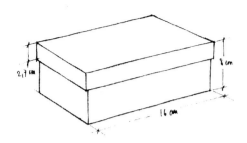

Start with a sketch of the box that you want to build. Decide on the dimensions and calculate the amount of material that will be needed. Grain direction is not yet important, but box and cover should be planned with the grain in the same direction. Some thought should be given to placing the pattern on the sheet with the least waste of material.

Pattern

Cutting

Draw the pattern for the box in a right angle to the edge of the board and cut out the entire rectangle (see the chapter on cutting). Draw all construction lines on the smoother side of the cardboard. Indicate sides and hinge flaps of about ½ inch.

Scoring

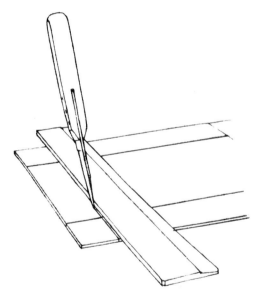

Score the four edges to a depth of two-thirds to three-quarters of the thickness of the board. Move the knife evenly and without stopping. Two or more strokes may be necessary.

If the score is not deep enough, bending may be difficult, and the board may crack. Scores that are too deep will result in weak joints.

Folding

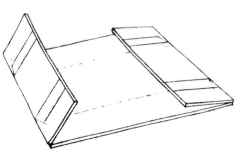

Fold up the four sides and push the bone folder along the newly formed edges, an often neglected but important step. The sides should not resist taking their new position, because some of the tension is removed by the bone folder. If this step is skipped, the sides will bulge out later.

Cutting out corners

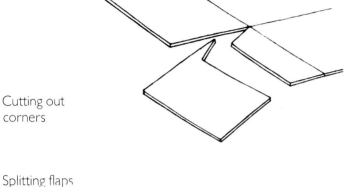

Cut out the corners using either a knife or scissors.

Splitting flaps

Hinge flaps should be as unobtrusive as possible. Split them by folding up the flap and forcing off the extra material. Fold up the sides to check that they are of equal height, and correct differences by refolding or trimming.

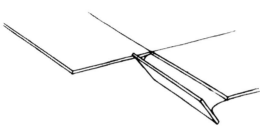

Gluing

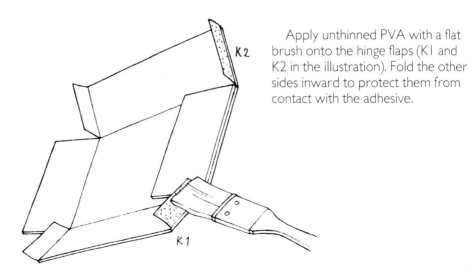

Apply unthinned PVA with a flat brush onto the hinge flaps (K1 and K2 in the illustration). Fold the other sides inward to protect them from contact with the adhesive.

Gluing

Rubbing down

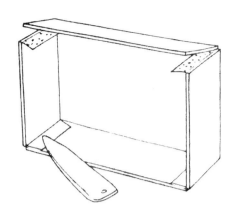

Attach the flaps to the adjoining side, adjust the position, and rub the parts together with the bone folder.

Gluing edges

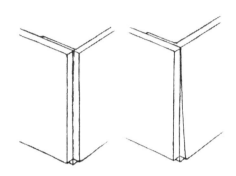

A correctly made joint is shown in the left drawing, a sloppy one in the right.

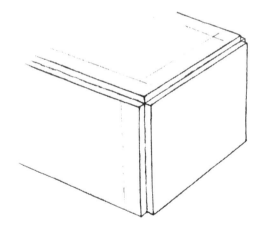

Drawing of a basic box construction.

Errors and corrections

The construction is now finished. Check it for flaws: Are all the corners right angles? Joints that have shifted can be repositioned, but if an inexact pattern is the cause of the problem, no correction is possible.

Sharp edges can be treated with fine sandpaper.

The cover

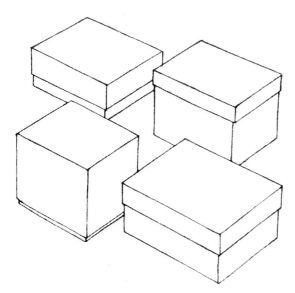

The construction of the cover corresponds to that of the box, except for slightly altered measurements. The rim, depending on taste and necessity, can vary in width. It may be as high as the box itself. Beginners are well advised to start out with proportions such as 1:3 or 1:4, since these seem to please the eye.

Fit

Measuring

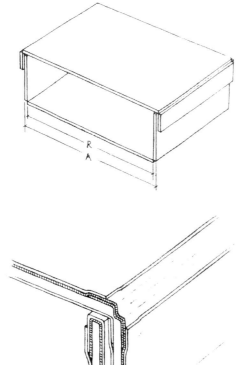

To find the right fit for the cover is somewhat tricky. There should be a slight resistance against opening once the box is covered and lined. Measurements for the cover have to be taken from the total outer dimension of the box (A), not from the measurement of the corresponding side on the drawing (R). About 1/16 in (1–2 mm) has to be added to the measurement of the cover.

This drawing shows how the thickness of the covering material and lining can influence the fit of the cover. On a vertical side, we are dealing with two edge-reinforcement layers, two layers of covering, and one lining. The quality of the craftsmanship obviously will have an influence on the fit.

EDGE REINFORCEMENT

A box constructed according to the preceding instructions could not stand up to real use — its edges are too weak. They have to be reinforced with strips cut from cloth, unless the box is entirely covered with paper or cloth.

Some boxes are merely reinforced; others are covered with paper afterwards, so that only small strips of the reinforcements remain visible.

If you are planning to cover the box, the colors of the reinforcements have to be coordinated with those of the covering materials.

The width of the reinforcements should be about ¾ in (2 cm) and they should be cut parallel to the selvage. Cut the first two strips the length of the shorter edges of the box, plus the length of the two adjoining verticals and overlaps of about ½ in (12–15 mm) each.

Set the strips out on newsprint and apply adhesive, using a 2:1 mixture of PVA and paste.

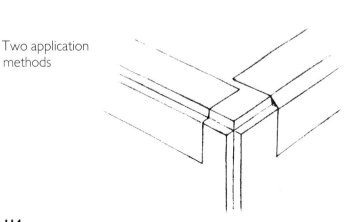

Strip width

Grain direction

Strip length

Two application methods

The strips can either be stretched over the edges or forced into the scores, which gives the finished box a more pronounced appearance.

Reinforcing the longer edges

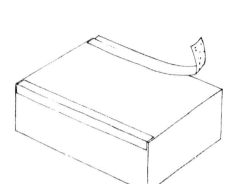

Glue the longer strips first. It is helpful to mark the position of the strips on the board, since the edge of a crooked reinforcement strip will be visible even through colored paper. Attach the strips in this sequence: Touch the strip to one side, rub it on gently, and fold it over to the other side.

The bone folder may leave shiny marks. To avoid this, place a layer of paper between it and the strip.
(Most of the drawings in this chapter are simplified in their presentation of edges for the sake of clarity.)

Other edges

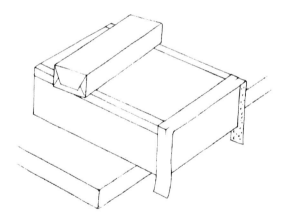

The other edges are reinforced as follows: Apply adhesive to the strip, touch it to the bottom edge, pull it around to the side, and rub it on.

Overlaps

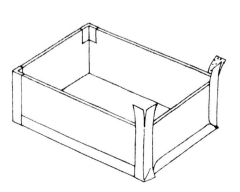

Make incisions at the corners, turn the strips tightly around the upper edges of the sides, and rub with the bone folder.

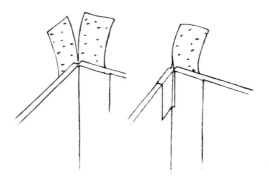

A detailed view of the upper corner.

Corner treatment

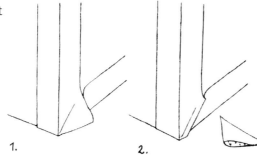

This sequence shows in four steps how the lower corners are treated. Note that the cuts do not reach the corner.

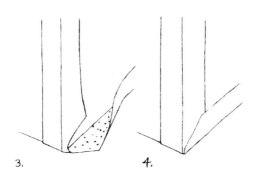

1. Push the strips together.
2. Cut off the extra triangle.
3. Lift the edges off as far as necessary to overlap.
4. Rub edges.

Finish

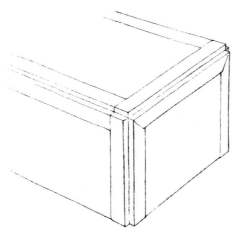

Rub all the edges and especially all the corners once more with the bone folder, to make sure that no spot has been overlooked. Check the box and cover in this way. The fit of the cover will of course depend on how carefully the components were made.

COVERING THE BOX

Each side of the box is treated separately. For the overlaps you need to allow about ½ in (12–15 mm) extra. Thin strips of the reinforcements will still be visible when the work is completed.

Visible edges

The drawing on the left shows a cutting pattern for the covering of a box and its cover.

Pattern for cover

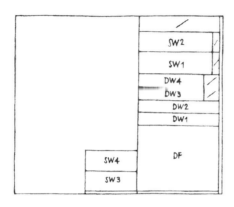

First the right edge of the sheet is trimmed, then a section that matches the longest dimension of the cover is cut off. It contains the parts DF (top of the cover) and DW 1–4 (sides of the cover).

Since there is some material left, covering material for two sections of the box itself can be cut from the strips: SW1 and 2 (side covering). The remaining pieces of covering material, SW3 and 4, have to be cut from a new section. The bottom of the box will be covered with a different material.

Grain direction

If the pattern is arranged so that the grain direction of the paper matches the horizontal lines, it will be easier to fold the overlaps. The covering for a box and its cover should always run in the same direction.

Expansion

The covering material will expand and contract under the influence of the adhesive. To achieve inside rims of the same width it is useful to test the material first.

Gluing

Setting on

Rubbing down

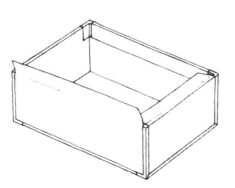

Using a 2:1 mixture of PVA and paste, apply the adhesive to one piece at a time. Position the covering material, rub it on, and fold it over the edges where necessary. If cloth is used, the rubbing has to be done with special care (see the chapter on adhesives).

117

Overlaps

Bottom

Covering the cover

Color

Lining patterns

With precise measuring, the overlaps of the covering material should match those of the reinforcement strips.

The bottom of the box is usually covered with a different and stronger material in a neutral shade that matches the rest.

As the last step, put the covering material on the cover.

LINING THE BOX

The lining counteracts the pull of the paper or cloth on the outside of the box. It covers everything but a small rim, along which the overlaps of the outside material remain visible. If the lining is omitted, the sides of the box will soon bulge out.

The lining is rarely exposed to view, but it should be installed with no less care than the covering material.

The lining paper should not resist being pushed into the corners, which limits its weight. Its color is not restricted to white: light-colored boxes especially benefit, often surprisingly, from the contrast that a dark interior provides. The colors of the future contents may provoke ideas.

The inner dimensions of the box provide the measurements for the lining. Add ¼–½ in (5–10 mm) to two opposing sides, for overlap at the corners. The height of the lining strips should be ⅛ in (2–3 mm) less than the height of the box. A narrower lining strip shows too much of the covering overlap and looks bulky.

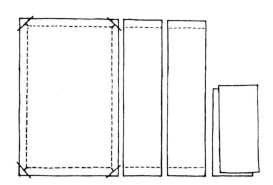

Lining the bottom

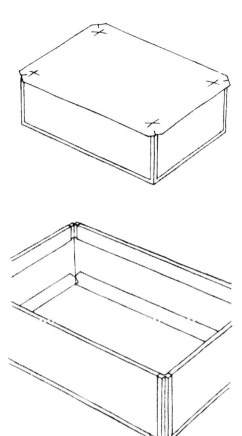

See how the four corners are beveled and cut. The cuts end exactly at the points at which the four corners of the bottom are located. The expansion of the paper should be considered as carefully as possible.

Apply adhesive to the bottom piece twice, place it on top of the box, and with four fingers push it down gently, taking care to keep it horizontal. If you used enough adhesive it should be easy to nudge the paper into the right position. Rub it down with movements radiating outward from the middle. The paper should lie flat with no wrinkles, especially in the corners (which may be a problem in deeper boxes), and no air bubbles should be trapped underneath. The answer to most difficulties is, again, the generous application of adhesive. The work cannot be saved if the paper sticks to the wrong spot at the wrong time.

To avoid adhesive stains at the upper edge of the box while the bottom lining is being inserted, you can finish this step before the outside is covered.

Apply adhesive twice to the linings for the two longer sides and insert them. The overlaps onto the narrow sides should have the same height as the rest of the strip. No beveled or rounded corners! The bottom edge should meet the bottom of the box.

Lining the sides

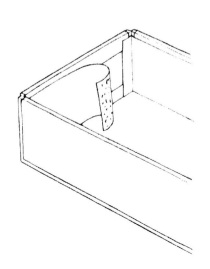

The last step is to install the lining for the shorter sides.

Even edges

Lining the cover

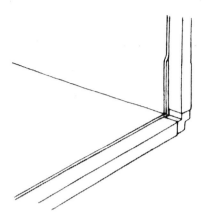

The general appearance of the work improves greatly if all parts that are visible at the upper edge look neat and even.

Line the cover in exactly the same way as the box. Be patient! Wait at least 2 hours before trying the cover for fit, otherwise, still-damp sections of lining may get scratched off.

With a large box, put blotting paper inside the box and cover and weight them down until dry. This prevents warping of big expanses of paper.

A BOX COMPLETELY COVERED WITH PAPER OR CLOTH

A completely covered box is not necessarily sturdier than one with edge reinforcements, but it has, without doubt, a different character (see the color illustrations on page 135).

For reasons of stability, only smaller boxes should be treated in this way, but the possibilities for color combinations are endless.

Fabric or paper is attached with a 1:1 or 1:2 mixture of PVA to paste.

See the chapters about cutting and gluing if necessary.

General considerations

Adhesives

Covering the box

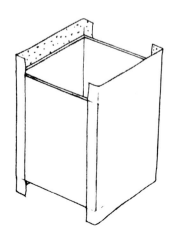

For the box itself you need five pieces of the covering material. Cut them from the sheet so that the grain direction is parallel to the upper edge of the box, to facilitate folding.

The first two parts overlap about ½ in (12 mm) on all sides. First fold the sides, then the bottom, and lastly the upper edges. Treat the corners as described on page 116.

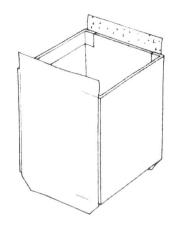

The other two pieces are 1/8 in (2–4 mm) narrower than the sides of the box, but they have top and bottom overlaps of the same width as the first pieces. The bottom corners are mitered at a 45-degree angle.

Cover the bottom of the box last, and use a piece that is 1/16–1/8 in (1–3 mm) smaller than the box bottom. Even though the bottom is rarely seen, the covering should match the other materials.

After all the overlaps are folded, rub all parts with a bone folder or fingers, then line the box as described on page 118.

Shallow covers

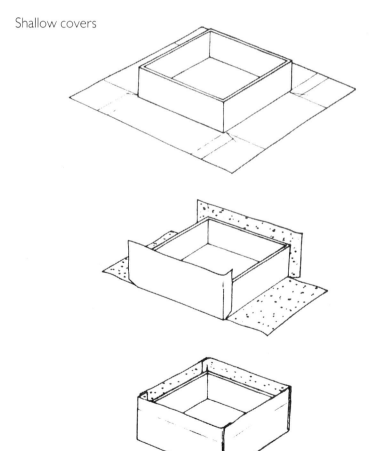

Coverings for shallow covers can be cut in one piece. The illustration shows corners already marked. Adhesive is applied to the cover, which is then set onto the paper.

Note that the lines marking the longest cuts are not a continuation of the sides of the cover, but are placed 1/16 in (1–2 mm) inward.

Cut the corners, paste the sides, and two opposite sides of the cover are completed. Do not forget to pinch in the paper around the corners after the overlaps are folded.

Now glue the other two sides and gently rub down the paper.

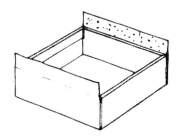

With scissors cut the corner sections and fold in the overlaps. The box is now ready for lining.

Deep covers

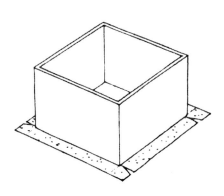

Deeper covers cannot be treated this way because it is not only a waste of material to cut the entire cover in one piece; it is also very cumbersome. Instead, cut a piece to the dimensions of the top of the cover plus ½ in (12 mm) on each side. Mark the position of the cover on it and apply adhesive to the entire piece. Place the cover onto it, rub on the paper, and cut out little triangles, as shown in the illustration. Remember that the points of the triangles are not exactly at the corners of the cover.

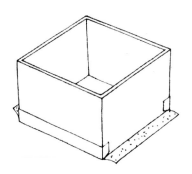

Do not forget to treat the corners correctly after two flaps are done before proceeding to finish the other sides.

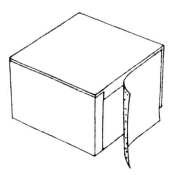

This drawing shows the cover from the top. Two sides, usually the longer ones, are already finished. The third and fourth are just being covered. After completion of this step, the cover can be lined.

THE HINGED COVER

Construction

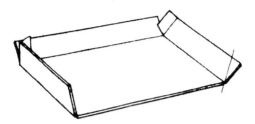

The construction of a cover with a hinge should present no difficulties if you have tried your hand at a regular cover. The drawing shows the differences:

One long side is missing. The adjacent shorter sides are mitered at 60 degrees.

The construction of a box with four fixed sides is described on pages 109–113.

Combinations

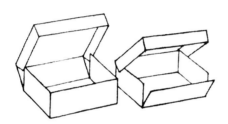

Other variations on the theme include a front that folds out, or a folding side attached to the hinged cover, which would be very similar to the hinged box that will be described later.

Reinforcements

Let us assume that all edges are already reinforced. Connecting a cover or a folding side to the body of the box is a similar process.

Hinge strips

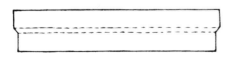

The outer and inner hinge strips have to be cut to fit the exact dimensions of box and cover. Remember to allow for shrinkage, since the strips are cut parallel to the selvage.

Attaching the cover

Apply PVA with a small amount of paste mixed into it and attach half of the strip to the cover.

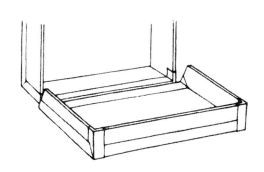

nner hinge strip

Now turn the cover upside down and attach the box. Stand the box on one side; its distance to the cover should be about one and a half to two times the thickness of the cardboard. Press it onto the fabric strip underneath, then close the cover and rub down the strip. Now open the box again and attach the inner hinge strip. Rub it down as described on page 126.

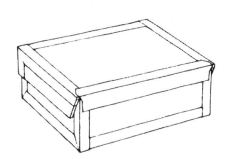

If the box is to be covered, this should be done after the cover is attached. The process is the same as for a box with a regular cover.

THE HINGED SIDE

Scored hinges

A hinged side is recommended if the contents of the box would otherwise be inaccessible without turning the box upside down. Often the hinged side is combined with a hinged cover.

The hinge can never be replaced by a mere score. If the parts are to be functional, covers and sides have to be cut separately.

Height of flap

Length of flap

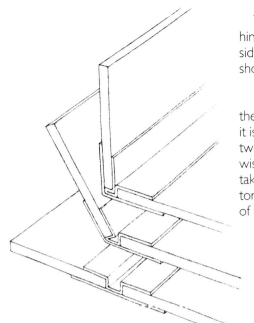

The drawing shows how the hinge works and why the hinged side has to be cut 1/16 in (1–2 mm) shorter than the other sides.

The length of the hinged side is the same as that of the entire box if it is supposed to rest against the two sides in closed position. Otherwise the length of the hinged side is taken from the length of the bottom, from inner edge to inner edge of the two sides.

Reinforcing box edges

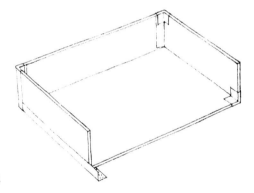

Reinforce the edges on the box while it is still missing a side. See pages 114–116 for details.

Connecting strips

Two strips of cloth will form the hinge. The outer one should be cut from the same material as the edge reinforcements, while the inner one should match the color of either the cardboard or the lining.

Cut the strips in selvage direction, with the necessary overlaps of 1/2 in (12–15 mm) each.

Set the side onto the strip and fold the overlaps.

Gluing the outer hinge strip

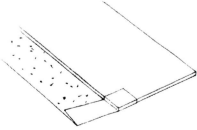

If a covering is planned, now is the time to cover the hinged side.

Connecting the flap

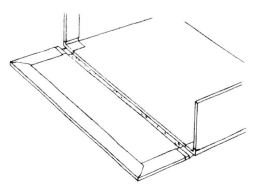

The part of the strip that is still free is covered with pure PVA. After a few minutes of drying time, set the box onto it. Remember to keep the edges at a distance of two times the cardboard thickness, and arrange the pieces so that the side edges are flush with each other. Turn the piece over and rub the strip carefully, but not until the area is supported with a block of wood to keep it from sagging.

Inner hinge strip

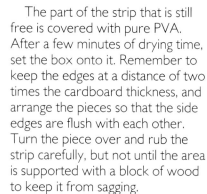

Apply adhesive to the inner hinge strip and set half of it down on the bottom of the box while holding the other half up to keep it from touching anything. Now press it gently into the score between the box bottom and the open side. No force should be used. If the bone folder does not slide freely, put a thin sheet of paper under it.

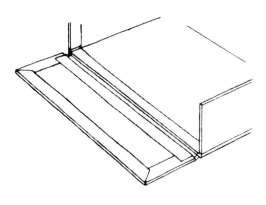

Then glue the rest of the strip onto the side. If a thinner cloth is used for the inside strip, the adhesive should contain more paste.

Lining the flap

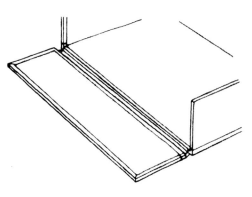

With a paper-covered box, the last step after lining would be the lining of the hinged flap. Allow the newly constructed hinge to rest for 15 minutes before it is moved.

If you are a perfectionist, you can cover the free edges of the hinged side. This must be done before the

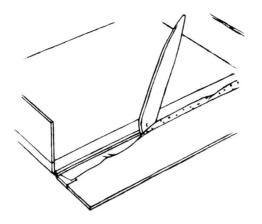

hinge is constructed. The little strips should be flush with the lower edge and have a ½ in (12–15 mm) overlap on top. The width is 1 in (2 cm).

PARTITIONS

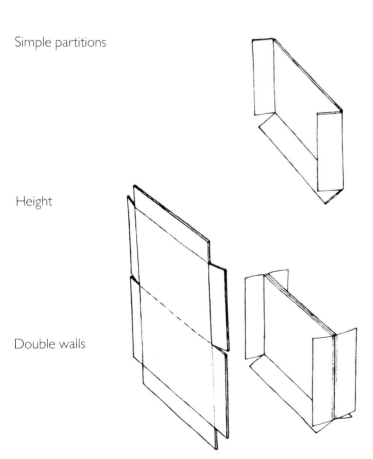

Simple partitions

Height

Double walls

Marking the position

Partition walls will keep order in a box that holds different items. Vertical walls can be positioned according to taste, but they have to be secured with folds. Simply gluing them in along three edges does not make an enduring construction.

Cut the walls large enough so that ½ in (1 cm) folds can be made along three sides, and split the folds as described on page 111 to make them less bulky. The grain direction of the partitions should be parallel to their horizontal edges.

The height should be either the same as the box itself or ⅛ in (2–3 mm) shorter.

If you want especially sturdy walls, construct them in double thickness. Connect the halves with pure PVA applied in dots and lines around the edges.

Weight and dry them for 20 minutes before inserting them in the box.

Mark the position of the partitions in the box before you attach them. Any deviation from right angles is not only aesthetically disturbing, but also diminishes the stability of the walls.

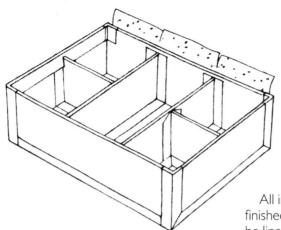

Snipping the overlaps

All interior walls have to be finished and dry before the box can be lined. Where a wall meets a side of the box, the overlap has to be clipped, but only to the spot where it touches the wall, not all the way to the edge of the box.

As before, when lining, start at the bottom and treat each compartment as if it were a box itself.

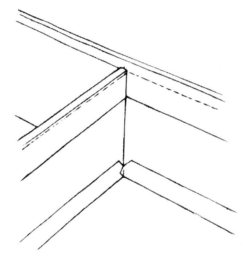

Lining the dividers

The upper edges of the compartment walls can be covered with 1 in (2–3 cm) strips of either the lining paper or the edge reinforcement material. After you line the sides of the walls, the box is completed.

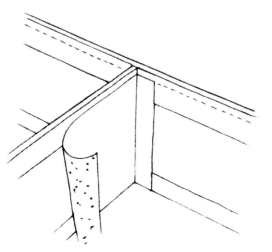

The drawing shows how the overlaps and linings should look. In this example, the partitions are the same height as the box.

THE PLATONIC SOLIDS

The color distribution of these six red, white, and blue tetrahedrons was inspired by their characteristic shape. Why paint them in the first place? For fun! To study relationships between shape and color, to explore regularities.

For best results the paper should be painted before it is folded, since it is then possible to let it dry under pressure.

Even though these nine cubes are painted in an unsystematic way, there is a definite relationship between structure and pattern. The color lines rarely follow the edges of the cubes.

The shape of the cube can be reinforced or counteracted by color. The obvious appeal of optical illusions should not distract from more serious approaches. Interesting changes can be observed when cubes of the same size are painted uniformly white, yellow, purple, or gray. To observe means to study and to participate intensely in the process.

On the upper cube the stripes that lead around the solid diagonally can be read as fifteen layers that make up the cube. The lower cube has two red corners, diagonally opposite each other, connected by red lines. Color can be introduced not only by painting but also by applying other papers. This, of course, makes countertension on the inside necessary, except when self-sticking materials are used.

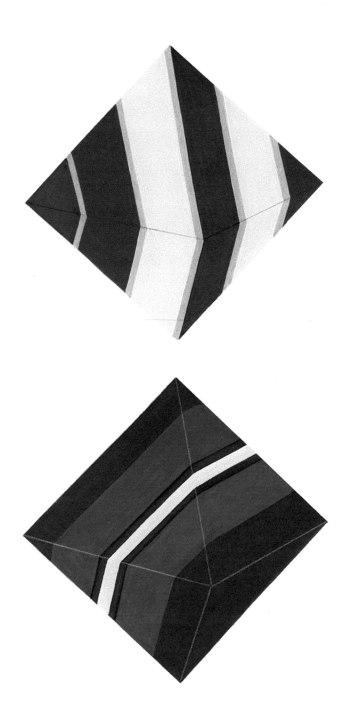

The stripes on the upper octahedron are parallel to the square that bisects it. The distribution of the stripes is asymmetrical.

On the lower octahedron the stripes follow the edge of the hexagon that bisects the octahedron, and gives it an entirely different appearance. An astonishing abundance of variations can be explored systematically or in a playful way.

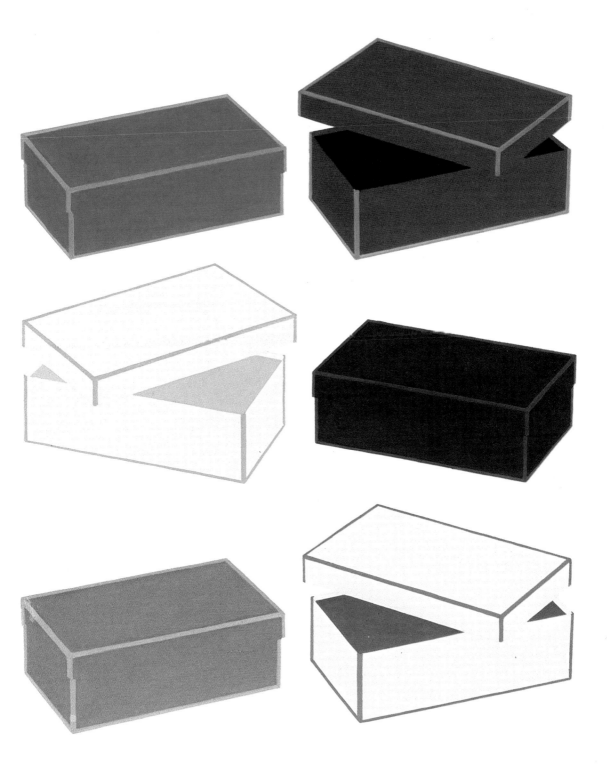

BOXES

The colors chosen for a box should always reflect its purpose. A toy box will naturally inspire a different choice than a tool box. Neither decision should be left to chance.

A cover may be covered with paper of a different color than the bottom of the box, as long as the two parts are treated as a unit. Comparable contrasts can exist between cover and lining. One box may be a volcano, cool and gray on the outside, glowing orange on the inside; another a yawning kitten, gray-brown, showing a brilliantly red mouth.

Six boxes entirely covered and lined with paper or cloth.

Two fabric-covered hinged boxes with labels. The label can have a colored rim or be a solid color. Even the writing on the label can provide a color contrast. In a balanced composition, the smallest detail influences the entire design.

Frequently a simple white label with black lettering will suffice.

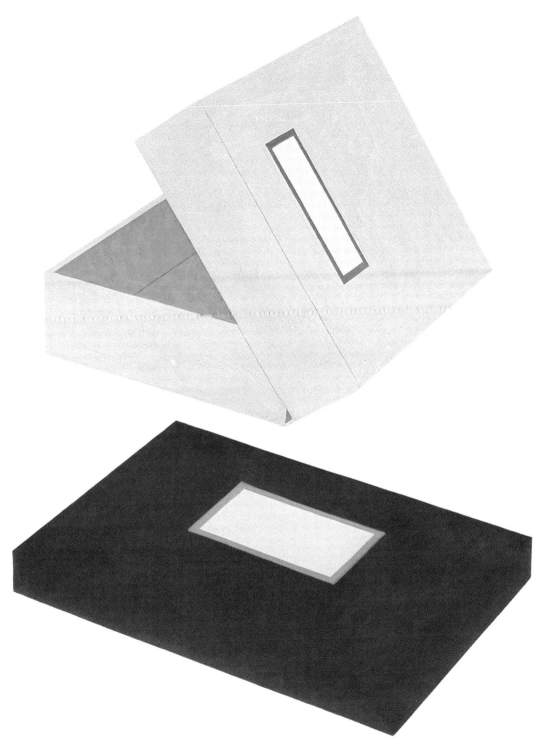

Two fabric-covered hinged boxes, decorated with labels on the front. Beyond the utilitarian purpose of such labels, they provide decorative accents.

If the contents of a box will be white, a white lining seems appropriate.

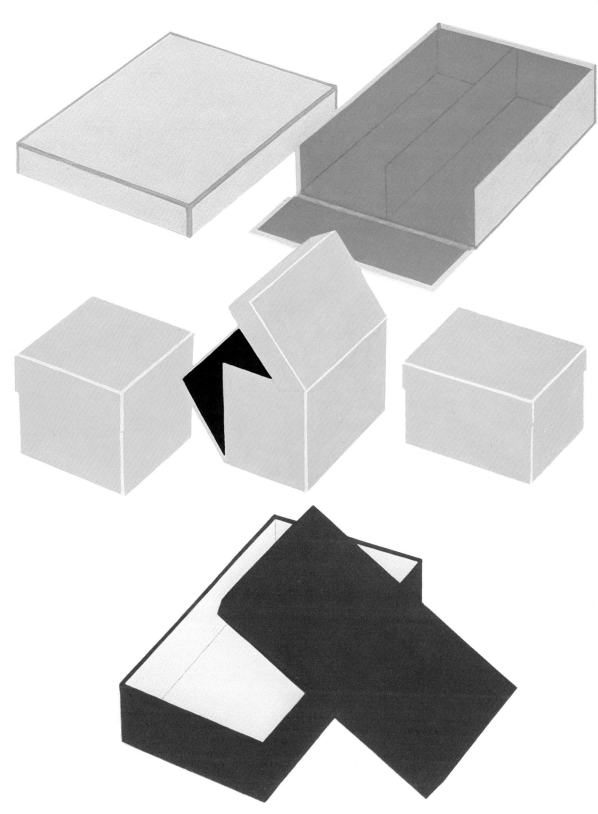

The shape and color of each box should form an aesthetic unit. As the top example shows, even gray covers can appear chromatic if the color of the edge reinforcements is chosen correctly.

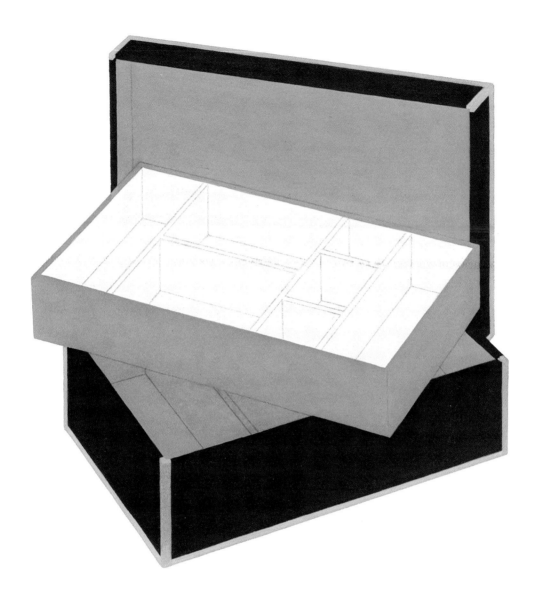

The combination of colors chosen for this box with an insert was inspired by a Japanese woodcut, or rather by my memory of it. Domestic scenes on such woodcuts often include little chests or boxes, and their coloration is, like the rest of the work, very appealing to most of us. But it is just as good, if not better, to find inspiration closer to home. A glance into a folder with colored papers might provide ideas.

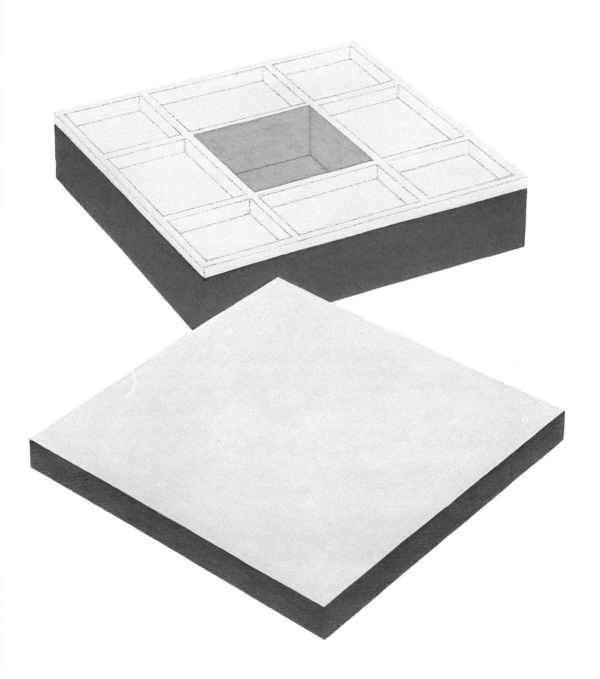

A square box with a cover and compartments for playing cards, dice, and other items. The lining and the upper edge were completed before the outside. The intensity and value of the red and blue are the same; the light gray provides the same contrast to both colors.

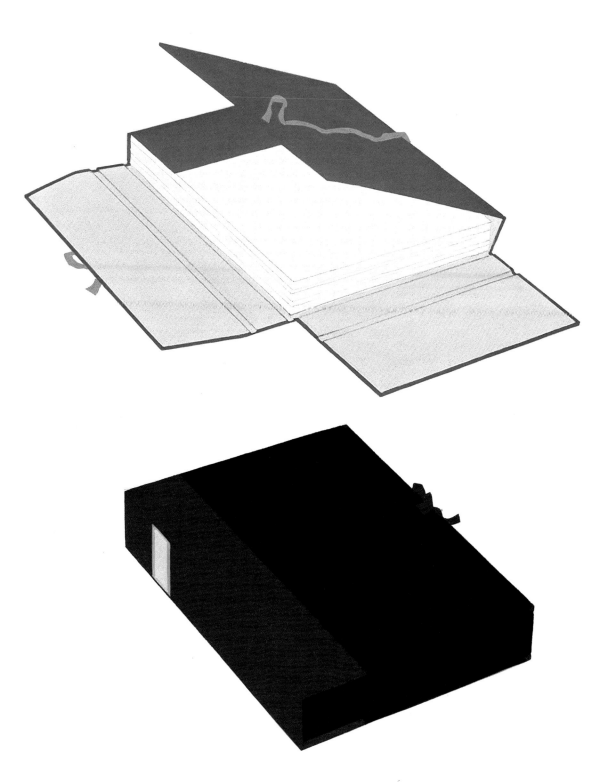

PORTFOLIOS

Two portfolios, each with three flaps. The upper one is entirely covered with cloth. On the lower one only the spine is cloth-covered; the rest is covered with paper. Any color, including gray and black, can appear lively against a contrasting accent.

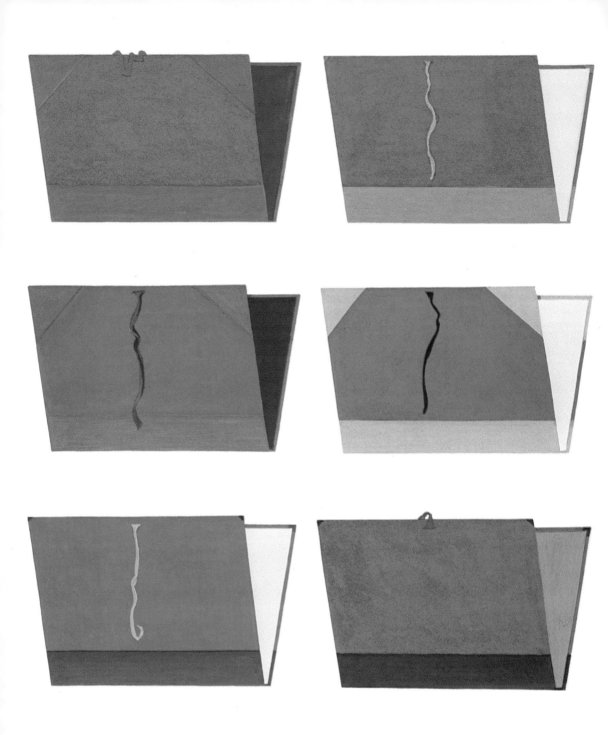

All six portfolios on this page have the same blue covering. They are differentiated by a variety of materials for back, corners, and lining. This juxtaposition shows how the adjacent colors change the character of a hue.

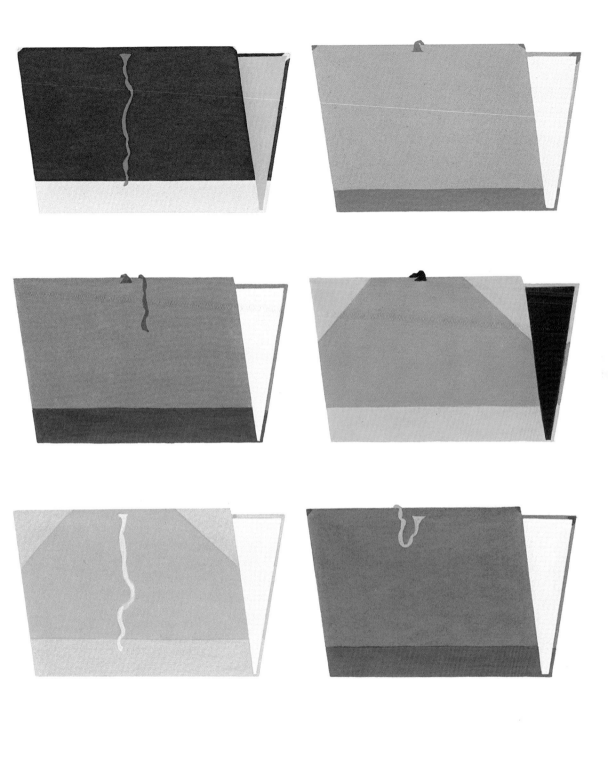

Six portfolios with cloth spines and paper covers. Large corners such as those on two of the folders will probably be a rare choice.

If store-bought papers do not offer enough variety, you can color or decorate selected sheets with the help of various instruction manuals or kits. I caution you against opulence.

Also consider the color of the ribbon; it is an element of the overall design.

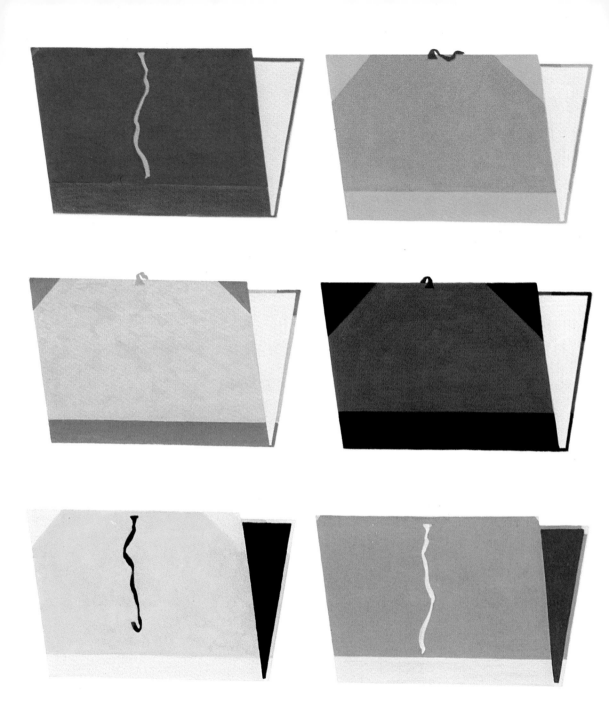

Six more portfolios with cloth spines and paper covers. Note the differences caused by varying the width of the spine: The size has to be chosen from an aesthetic rather than from a practical point of view. More on the subject can be found in the text.

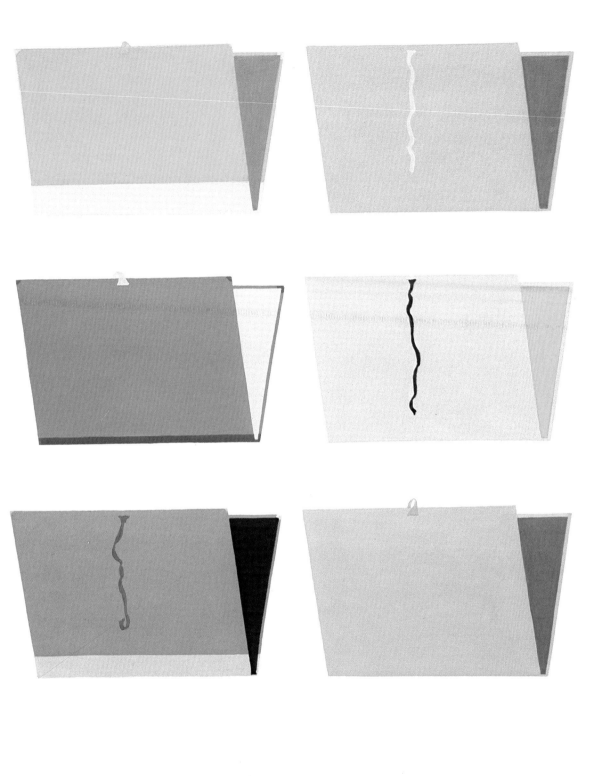

Three simple portfolios with cloth spines and very small corners on the left, three portfolios completely covered in cloth on the right. If the choice is between a colorful ribbon that might not be strong enough and a plain one that seems sturdy, choose the functional one.

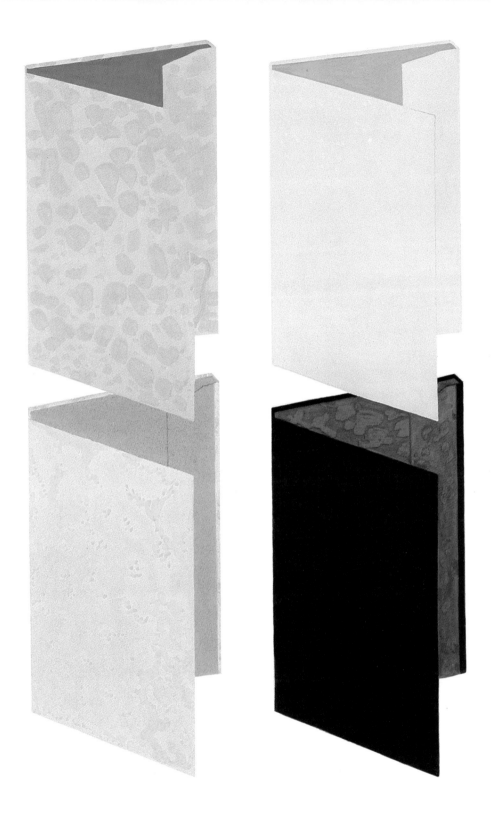

Four cloth-covered portfolios with side flaps. All spines, edges, and corners are reinforced with cloth.

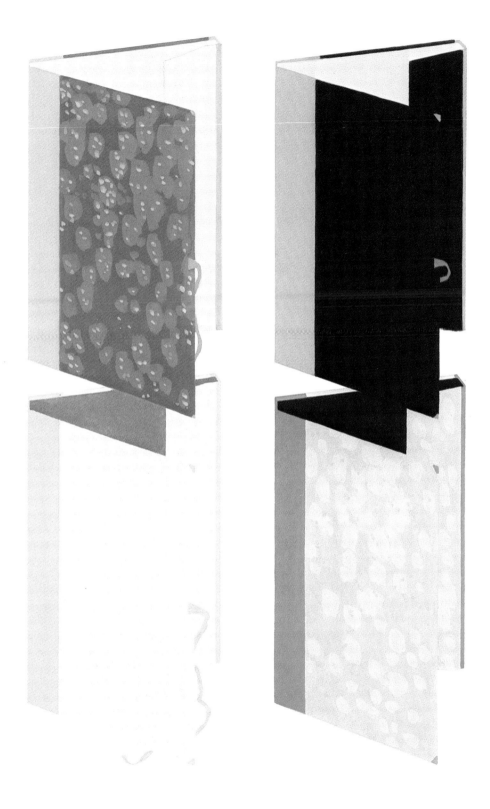

Four portfolios with side flaps. The flap of the folder at top left is covered with the lining material; the flaps of the other three have flaps covered with the same material as the outside. Cloth can be marbleized just like paper for use on boxes, portfolios, and books.

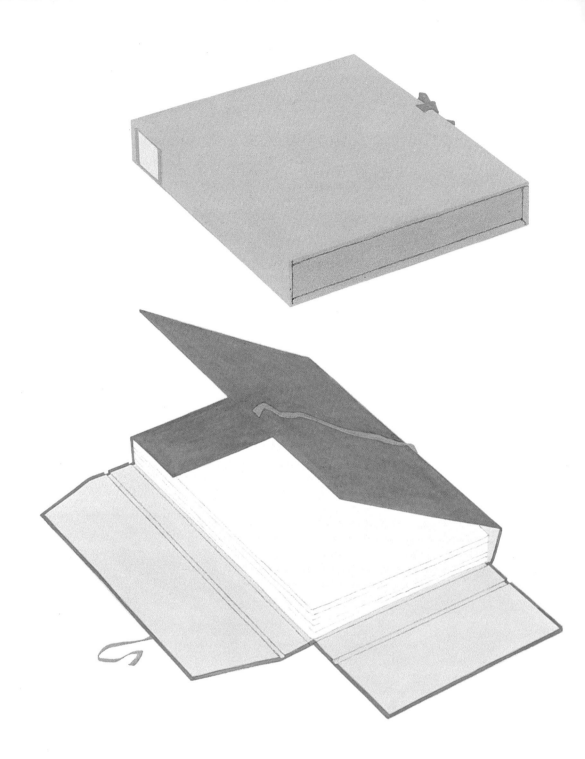

Two cloth-covered portfolios with three flaps. For a discussion of labels and lettering for books and portfolios, see pages 284–287.

BOOKS

Six cloth-covered books. The colored edges are superfluous, but like headbands, they are ways of planning the total image of the book. Colored edges are, however, not necessary, and a balanced design is possible without them.

Twelve books with cloth-covered spines. Headbands are included in the illustrations. They are one of the factors that help a cover come alive, even without the aid of questionable decorations.

There is a practical reason for the use of colorful papers: Spots and stains may be less obvious. Traces of use should never be considered detrimental to a useful object, but abuse should never be evident.

Simple covering and cloth spines, formerly called half bindings, can be beautiful. The care invested in the work and the choice of materials is a necessary ingredient for success.

Labels are not usually seen on books, but in the absence of printing equipment they can be an appropriate way of lettering.

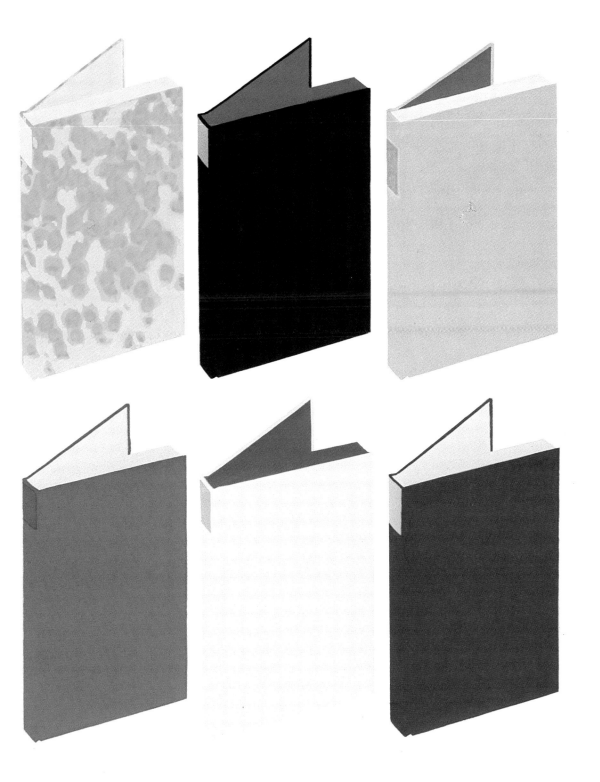

Limitations in typography and in colors for cover, block edges, and labels can prove inspiring indeed. Freedom without boundaries is not a prerequisite for imagination and variety of thought.

Two cloth-covered books or albums. Color design should be not an afterthought but an integral contribution to the total plan. The more the work is approached as a unit, the better are its chances for turning out well. Sensational effects are all too often pursued, to the detriment of "just right" solutions. The extraordinary cannot be forced into existence, but it might result unexpectedly and surprisingly from constant and earnest effort.

Six cloth-bound albums or scrapbooks in different formats. They are bound with thread or cords using various types of stitching.

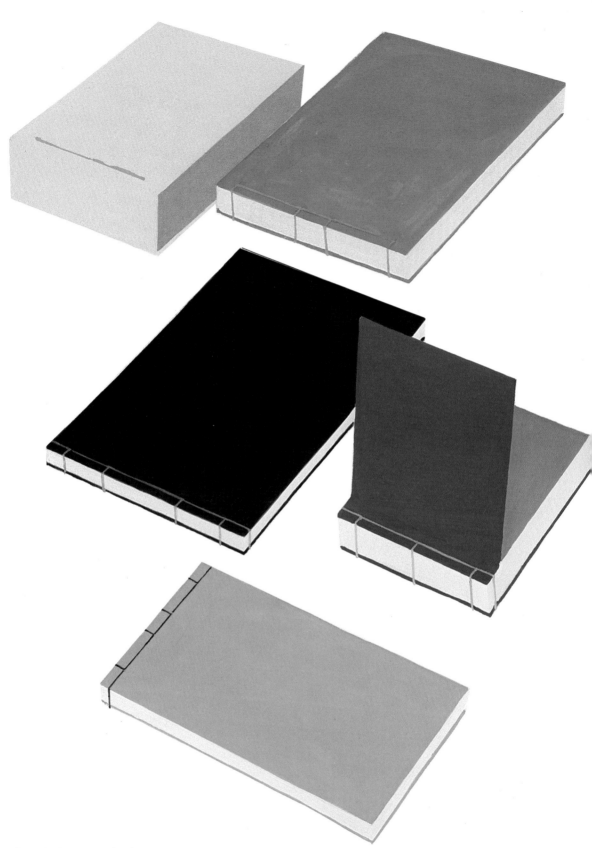

Five cloth-covered side-sewn albums.

INSERTS

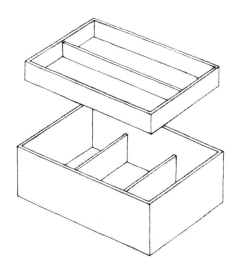

Insert and box

An insert is a box in a box. It is, of course, constructed like a box. It can be covered and lined with the same paper used to line the box. Whether or not the edges should be treated with reinforcement strips has to be decided for each individual case.

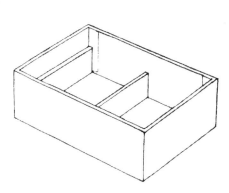

Supports

If the box has only one or no dividing wall, edge supports are needed, on either two or all four sides, depending on the size of the box. This complicates the lining process. Cuts have to be placed to make a pattern before the actual piece is glued into position.

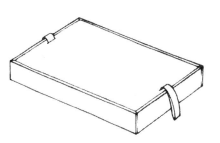

Handles

If no insert wall offers itself as a handle, attach two little loops before the insert is lined, but after it is covered. They are guided through slits in the bottom of the insert and glued in place.

THE HINGED BOX

There is a close relationship between boxes with hinged sides and hinged boxes. Imagine that the cover of a hinged box is attached to the movable side, and that the sides of the cover are the same height as the sides of the box.

Many things can be conveniently stored inside: notebooks, photographs, postcards, mats, and so on. Items can easily be removed or shifted from box into cover, and, closed, the box can be stored upright on a shelf like a book.

The construction of such a hinged box should be attempted only after some practice with easier projects, because the steps that have been explained elsewhere are not repeated here.

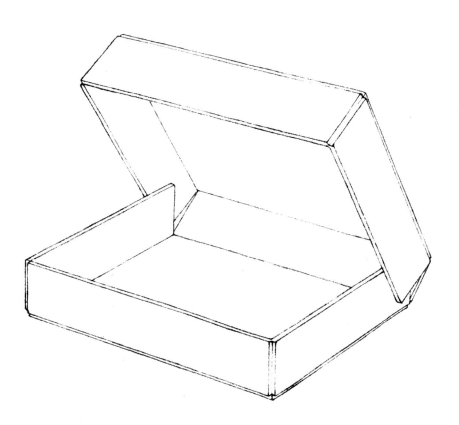

Side views

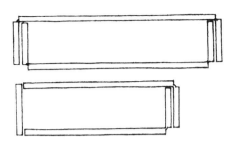

The two different views at left show how the box works. Bottom and cover are, except for their sizes, exactly the same, and their sides are the same height.

See page 113 for information on fitting bottom and cover together.

Rectangular or trapezoidal spine

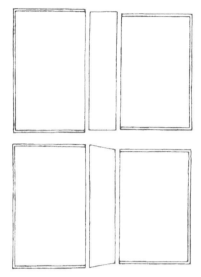

These two drawings show variations in the spine of the box. Both are functional, but the rectangular spine is preferable if the box will be stored on a shelf.

A BOX WITH CLOTH SPINE

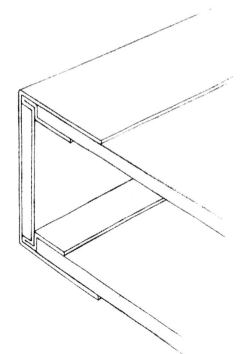

CONSTRUCTION

Detail of spine

Depending on the kind of cloth used, the back side or spine has to be about $1/16$ in (1–2 mm) lower than the other sides (see page 125). Otherwise the construction is the same as that of a box with a hinged side.

159

Construction

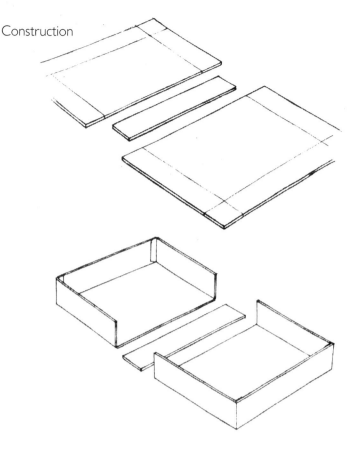

The length of the spine matches that of the cover. (In the case of the trapezoidal spine, the shorter edge corresponds to the bottom of the box.)

The drawing shows the three components before the sides of the box and cover are folded up. The exact measurements for the cover can be obtained only after the bottom has been constructed. See the chapter on boxes with covers for details.

The sides are folded up and glued. Then the edges are reinforced, as in the hinged box.

EDGE REINFORCEMENTS AND HINGES

Covering the spine

Connecting the parts

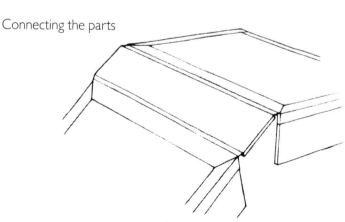

The drawing shows a spine that is already covered. The overlaps at the ends, shown folded over, are slightly beveled. The side overlaps are ¾–1 in (2–3 cm) wide. Use the same cloth for the spine as for the edge reinforcements.

Apply unthinned PVA to the two folds and attach the box and the cover at a distance of one and a half times the cardboard thickness. Rub from the outside.

The distance of both parts to the spine has to be exactly equal. Weight the pieces and let them dry for ½ hour.

Inner strip

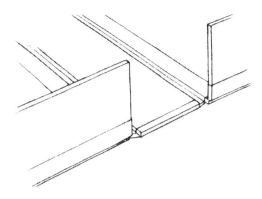

Cover

Lining

The last step in making a hinged box with reinforced edges is the application of the inner hinge strip. It strengthens the joint and provides countertension for the outer hinge strip. Cut it to its final dimensions before gluing it in. It is slightly longer on one long side than on the other because the cover of the box is slightly larger than the bottom of the box. First glue the strip to the inside of the spine. Glue it to the cover and bottom only after working it into the grooves between them (see the drawing on page 184).

If you are not planning to treat the box any further, make the inner strip of the same material as the outer one. But if covering and lining are to be added, the inner strip should match the lining.

Possible covering materials and linings are described in detail on pages 117–120.

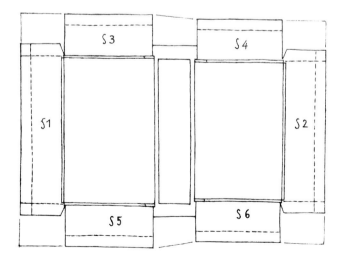

Cover pattern

A HINGED BOX COVERED WITH CLOTH

This project requires considerable proficiency, and should only be attempted after you have made a hinged box with a cloth spine and cover, as well as a box with a cover completely covered with cloth.

The drawing shows the three parts of the box laid out on the precut cloth. The construction of the box itself has already been explained.

Selvage

Corners

Expansion

Marking the position
of the cuts

The selvage of the cloth should run parallel to the spine; the cloth is one whole rectangle. Bookbinders cut off corners and other leftover parts after the adhesive has been applied to the whole piece and the cardboard pieces have been set onto the cloth, because the reaction of cloth to adhesive is often unpredictable. It may stretch across the width and shrink in length up to as much as an inch per piece cut for our purposes. It is therefore a good idea to mark the cuts only lightly with a pencil.

It is best to allow ½ to 1 in (1–2 cm) in selvage direction for shrinkage. Cutting of excess cloth is easy — to work with overlaps of less than ⅛ in (2 mm) is not.

Covering the edges

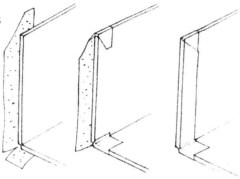

Before covering the sides, the four free edges next to the spine have to be covered. The edges of these strips will probably show through paper covering applied later, but you can diminish the effect by rubbing them down with the bone folder.

Applying adhesive

It may be difficult for an inexperienced person to apply the adhesive evenly onto a large area of cloth, but it is also possible to use a PVA-to-paste mixture of 4:1 instead of 1:1, and apply it to the cardboard.

Setting on

Cutting

Apply adhesive to one piece after another and set each onto the fabric. Do not let any part of the cloth sag or hang, though; support it with large wooden boards if necessary.

Now you are at leisure to cut slowly and carefully wherever indicated.

Folding overlaps

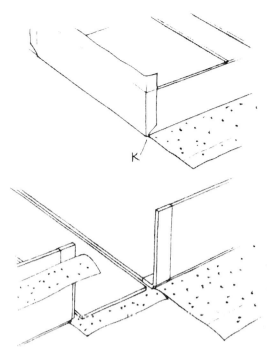

Glue sides S1 and S2 of the cover (see page 161) with a 1:1 mixture of PVA and paste. Do not forget to pinch around the four corners (K).

Next work on the sides (S3, 4, 5, and 6) and the overlaps at the ends of the spine, which is probably the most difficult step in this project. The cloth has to be worked into the joints with utmost caution. A very narrow strip of the overlap comes to lie underneath the edge of the box or the cover respectively.

Trimming overlaps

Inner strips

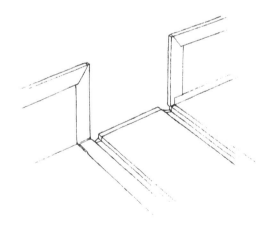

Lining

After this job is done, fold the remaining edges. Check, after making the cuts at the corners, to see if the adhesive has dried, and reapply if necessary. This is very important.

Unequal overlaps can be adjusted with the help of a straightedge and a sharp knife.

Start the lining of the hinged box with the cloth strip for the inside of the spine (see the drawing on page 183).

The material should match the color of the lining and has to be worked into the grooves very carefully. The strip exposes an area of about 1/8 in (2–3 mm) on each of the narrow sides of the spine. On the long side it overlaps 1/2 in (12 mm) as usual onto the edge of the box or the cover, and will later be overlapped by the lining paper.

The lining pieces are set down exactly on these hinge edges. The rest is executed according to the instructions on pages 118–120. The illustration shows the bottom lining already in place.

A ROUND BOX WITH A COVER

Constructing a round box is an entirely different project from making a rectangular one. It is considerably more demanding technically.

Start with a sketch, noting the desired dimensions. Avoid extreme sizes for a first try—a diameter of about 3 in (8 cm) and a height that is easy to line seem best.

The drawing shows the basic construction: two parallel circles of cardboard with a rectangular wall wrapped around them. The upper circle is reinforced by a second one.

Heavy covering and lining materials are even more difficult to handle in a round box than in a rectangular one. Otherwise, the same principles apply for the choice of paper or cloth.

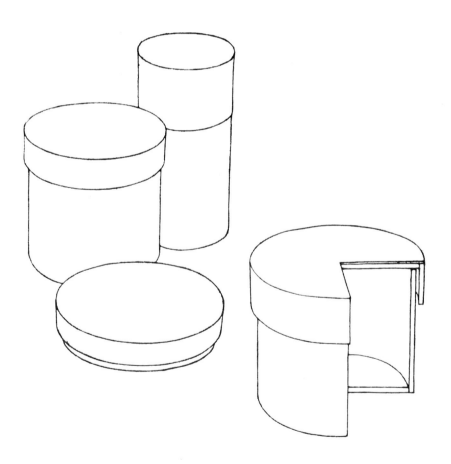

CONSTRUCTION

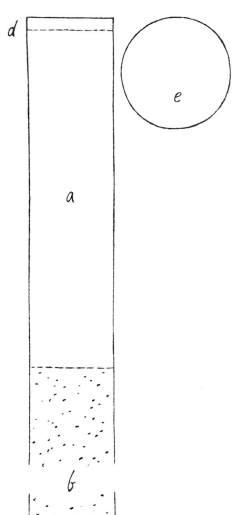

For the side-wall and bottom:
  a. The innermost layer. It will not come in contact with the adhesive and is protected during gluing.
  b. The second layer. The interruption indicated in this section means that it could be cut long enough to wrap around twice.
  c. End section. This is filed to a beveled edge, which will be positioned on the outside.
  d. This is the beginning of the strip and it is beveled on the inside.
  e. Bottom. This consists of two or three layers.

Gray cardboard is best suited to the construction of the box wall of round boxes. For boxes with a diameter of less than 4 in (10 cm) you can also use two-ply mat board or similar material. Avoid any brittle material that has a tendency to crack. The cardboard may also resist being rolled if the strip for the wall is cut with the grain. For the round top and bottom, a wider choice of materials is possible.

Cutting

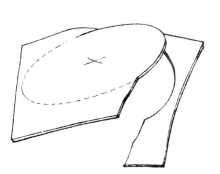

Material for bottom and cover

The bottom determines the interior diameter of the box, because the wall, as one long piece, is wrapped around it.

It is not a good idea to cut the bottom in one piece, because it should be about 1/8 in (2–3 mm) thick. It is better to cut two layers about 1/16 in (1–1 1/2 mm) thick and join them.

165

Gluing

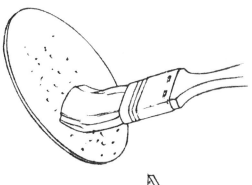

Glue the bottom together with pure PVA, applied to both sides, and let it dry for 2 hours.

Sanding

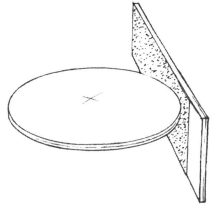

Smooth the edges with a sandpaper file. To make sure that no more than necessary is removed from the edges, draw a circle with a slightly smaller diameter onto the cardboard as a guide. Carefully remove any ridges that build up during filing.

Cutting the box wall

For a box diameter of 3 in (8 cm), aim for a wall thickness of about 1/16 in (0.7 mm) consisting of three layers. The box will be most stable and circular if the wall is wrapped from one continuous strip rather than from several connected ones. If connections are necessary, the edges have to be beveled, glued together with pure PVA, and weighted for 15 minutes.

Rolling the box wall

Length of the wall

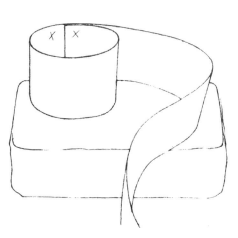

Cut the wall strip to the right width, wrap it three times around the bottom, and mark the beginning of the first and the end of the third turn along the edges with a pencil. This is not, however, where the cuts will be placed. Add ½ in (15 mm) on each end.

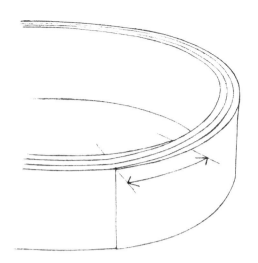

A reminder about grain direction: If the strip can be cut across the grain, there will be less difficulty with shrinking, wrapping it will be easier, and the finished wall will be sturdier.

The drawing shows how beveling the ends makes an even thickness possible.

Matching the ends

Beveling the edges

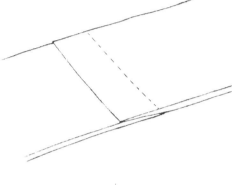

To bevel the ends, use a sharp knife and sandpaper. To check your progress, hold the ends against each other. You have achieved an ideal result if you cannot detect any difference in thickness when your finger glides over the joining place. The evenness of the roundness depends on your success in this step.

Gluing

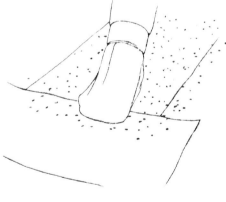

Place the strip on newsprint with the inner edge facing you and the inner surface facing up. Cover the edge up to the pencil line that marks the end of the first wrap and apply PVA.

Next, apply PVA to the edge of the cardboard circle, either directly or by rolling it over the adhesive.

167

Wrapping the wall

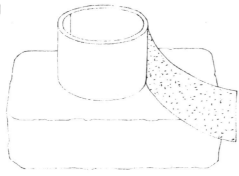

The final wrapping of the wall has to be done swiftly. The top of the box should be identical to the bottom in size, so hold the strip close to the bottom. Do not wrap loosely or trap any air bubbles. Make sure that the bottom is flush with the wall.

Sanding the edges

After ½ hour of drying time, sand the upper and lower edges and correct any irregularities at the ends of the strip with a sandpaper file.

Cover construction

The cover of the box can go part-way or all of the way down the box. To achieve the right fit, you must take into consideration all the covering and lining materials, including their double thickness at overlaps. If the work is not done carefully, folds and wrinkles may interfere with a good fit.

Determine the diameter of the cover, cut a circle, and proceed to make the wall as before.

As an additional reinforcement, a second circle can be superimposed on the first one. Cut it about 1/16 in (1 mm) larger than the first.

Cover reinforcement

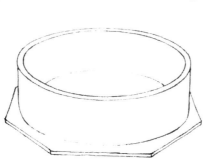

Apply PVA and weight the cover for ½ hour. Then cut any protruding rim off with a pair of scissors and sand the edge.

To keep an even tension, place another reinforcement disk on the inside of the cover. This also improves the stability of the cover.

Another advantage of this construction is that the top of the cover, which usually is seen first, will stay smooth all across its surface.

COVERING THE ROUND BOX

Edge reinforcements cut from cloth may seem heavy on a round box, even when it is also covered with paper. All-paper or all-cloth coverings are best. Besides, round boxes are generally stable enough and do not need the extra reinforcement.

To cover the wall of the box, cut a rectangular piece of cloth with an additional ½ in (12 mm) on each side. The length of the strip should preferably be parallel to the selvage. If you use paper, the grain direction should be parallel to the top and bottom edges of the box. Allow for a ¼ in (6 mm) overlap where the covering joins.

Use a 1:1 mixture of PVA and paste. Set the material down along a pencil guideline. Be sure that the overlap is equal on top and bottom, and that you obtain perfect adhesion.

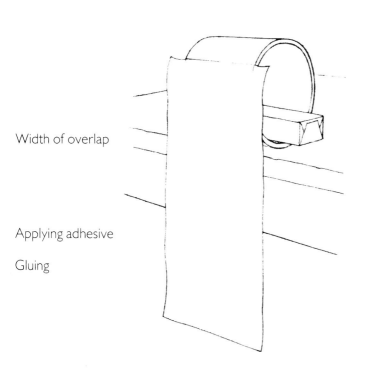

Edge reinforcement

Width of overlap

Applying adhesive

Gluing

Cuts

Folding

With a pair of scissors make a series of parallel cuts about ¼ in (6 mm) apart, closer in the case of a small box. The cuts should not quite reach the edge of the box.

Wrap the flaps over the edge, taking care not to tear off any of them, especially when you rub them with the bone folder.

Check to see if the adhesive has dried before turning the flaps inward, and reapply it if necessary.

Covering the bottom

Cut the round covering material for the bottom about ⅛ in (4 mm) smaller than the bottom itself. The paper should be sturdy and matched to the rest of the covering.

Covering the top of the cover

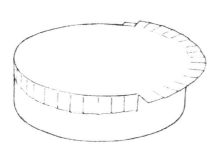

When covering the cover, apply the round top first.

Cut it with an overlap of ½ in (10 mm), apply adhesive, set on the cover, rub it, make cuts around the overlap, and fold them down. Protect delicate material from the bone folder with a layer of paper.

Covering the side of the cover

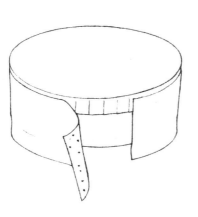

The covering material for the side should not quite reach the top edge. Fold the overlap in on the bottom, about ¼ in (6 mm), as before.

LINING THE ROUND BOX

Lining the bottom

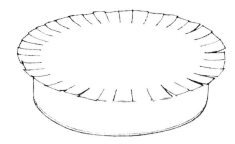

Lining the wall

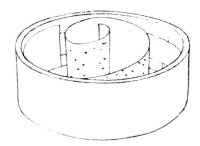

Start with the cover, because it is usually lower and easier to handle.

Cut the round lining part with a ¼ in (5 mm) overlap, which will eventually rest on the side wall of the cover. (For a box with a diameter much greater than 3 in (8 cm), you will need a wider overlap.) Make short, parallel cuts, about ¼ in (6 mm) apart, around the overlap. Apply adhesive twice, place the lining over the inverted cover, and push it in gently with four fingers. The lining should stay horizontal at all times.

If necessary, center the lining and rub it down with radiating movements, starting from the middle. None of the small cuts should be visible on the circle: They will all later be covered by the lining of the wall.

Cut the lining for the side of the cover with the grain of the material, and add ¼ in (5 mm) for the overlap. It should reach from the bottom of the cover to the edge. Apply adhesive, roll the strip up loosely so the layers do not touch, insert it into the inverted cover, and rub it against the wall of the cover.

During this process it is important to push the paper together rather than stretch it, because it will contract during drying and could easily separate from the wall, especially if not enough adhesive was used, or if the diameter of the box is very small.

The box itself is lined in the same way as the cover.

Long and narrow cylinders have to be lined before they are rolled together.

A ROUND BOX WITH A LIP

It is advisable to complete a regular round box with a cover before undertaking a box with a lip, because the steps of the process are explained in detail for that project. Only the new features are described here.

It is characteristic for a round box with a lip to have a cover of the same diameter as the box itself. The cover is held in place by the lip. This construction makes it possible for the contents to protrude from the opened box and to be easily accessible. It makes an ideal storage container for pencils.

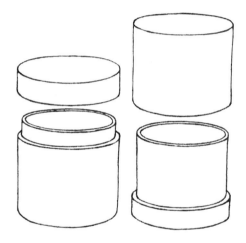

Two boxes with the same volume but different proportions.

CONSTRUCTION

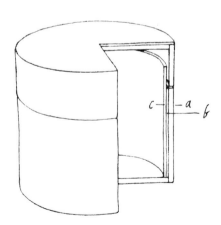

The drawing shows the construction of this box. For the sake of clarity, the wall (*a*) is represented by only one layer. A second layer (*b*) is inserted between the wall and the lip to give a little extra space, which makes opening and closing the box easy. On the inside you can see the lip (*c*), which supports the cover. When the box is closed, the lip should not quite reach the top of the cover, and so some free space will remain.

The circles cut for the bottom and the cover are the same size. The construction of the two main parts of the box is the same as for the box without a lip.

Gluing the lip

Covering

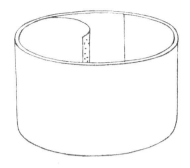

When you construct the bottom and the cover, put the middle layer in place. Glue a rolled-up strip of heavy (240 gm^2) cardboard to the inside of the box. It should be flush with the bottom and upper edge of the box. Its ends meet but do not overlap. Now box and cover are ready for covering (see page 169).

THE LIP

The lip is assembled before it is inserted into the box. The new inner diameter determines the outer diameter of the lip. For a box of 3 in (8 cm) diameter, roll a strip of heavy (300 gm^2) cardboard into three layers. The thickness of the lip should be about one half or two thirds the thickness of the box.

Measurements

Thickness of cardboard

Length of strip

Rolling

Inserting the lip

Cut the strip into the desired width, roll it, and place it inside the box, but do not glue it yet. First, mark the end of the first layer and of the whole strip with a pencil as before (see page 166), then cut, bevel, and sand the edges.

Cover the first layer of the lip with a protective sheet and apply pure PVA to the rest of the strip. Align the beginning edge with the corresponding line, and form the lip. After ½ hour the lip can be inserted into the box. If everything fits, sand all edges.

Covering the lip

Cover the visible section of the lip with a strip that reaches ¼ in (5 mm) below the upper edge of the lip. Snip the overlap and turn it inward at the lip. The paper should be the same as for the lining.

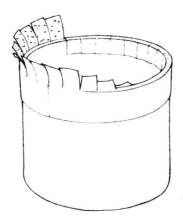

Lining the lip and bottom of the box

The lining almost reaches the upper edge of the lip. At the bottom edge, turn it to the outside. While it dries, line the bottom of the box. The overlap onto the side wall has to be rubbed down with special care, so that the lip will not get caught on the edge when it is inserted.

Inserting the lip

Insert the lip after 20 minutes of drying time. Apply pure PVA in dots along the edges of the inner box wall and take care that none of it is visible on the outside. Lining the cover is done exactly as described before.

PORTFOLIOS

Technically, any folded piece of paper could be called a portfolio or, more simply, a folder. Imagine making a second, parallel fold, using more substantial material, and adding strings or ribbons for closures, and there is something like the history of the portfolio.

Another prototype is a stack of papers protected by a wooden board or piece of cardboard on each side and tied with a string or a leather strap. It must have been the next logical step to connect the boards with thread or leather strips to form a hinge, or to loop longer ties through slits, as shown in the drawing below.

Another improvement would have been to replace leather strips with woven ties, which could be cut in any desired length to stretch across both boards. Folds could be added to protect valuable sheets inside. Ties served as closures.

These early forms are simple but useful, and their appearance has hardly changed over the course of the centuries. What distinguishes them from one another and makes them more than "primitive" is their individual style. Some fascinate with utilitarian form that seems unintentionally beautiful, others delight with joyful ornamentation without danger to the sovereignty of the overall design. I remember as examples a portfolio from eighteenth-century Japan, and a loose-leaf book from Africa dating from the same period. Closer to home we find images of portfolios in paintings by Jean Chardin, Honoré Daumier, and Wilhelm Leibl, all of them simple and convincing.

For color illustrations of a variety of portfolios, see pages 141–148.

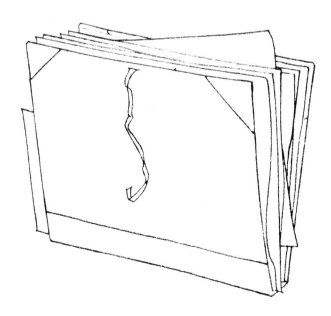

Let us start with some words that describe the parts of a portfolio. "Top" and "bottom" always refers to the portfolio as drawn in the illustration below, that is, in a vertical position, even though it will in reality be horizontal most of the time.

The first six terms are also used for books:
  a. front cover
  b. back cover
  c. back or spine
  d. front, or front edges
  e. top, or top edges
  f. bottom, or bottom edges
  g. front flap
  h. top flap
  i. bottom flap
  k. sides

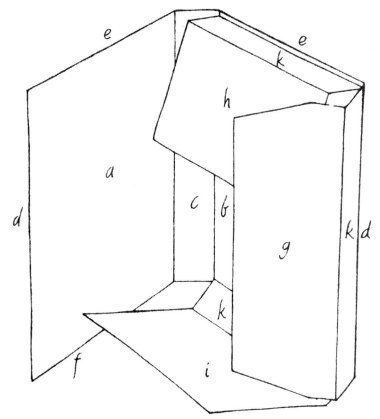

Among the many possible forms we will choose some that are most useful and easy to construct, but of course all of the examples can be varied to fit different demands.

The future contents have to be considered. There is a difference if a portfolio has to accommodate a changing number of sheets or notebooks, or if it is to protect a stack of fixed height. The first case requires a flexible spine, the second a stiff one. The intended use will call for a more or less sturdy portfolio.

If, for example, a portfolio is used to hold drawings during outdoor sessions it should be rather stable, so that it can double as an easel. It will also need flaps to protect the contents in the event of rain.

Size

The size of the portfolio also depends on its purpose. Will flaps be necessary? Should the portfolio be larger than the items it contains? How tightly should they be held in place?

Making a simple portfolio, partially or entirely covered with cloth, can be considered an exercise in bookbinding, since many steps are identical.

Grain direction

The portfolio should be cut so that its spine runs in the direction of the grain. The grain of the flaps should be parallel to their hinges.

CREASED AND SCORED PORTFOLIOS

Creasing or scoring?

Both creasing and scoring are possible, but creasing is preferable since it does not weaken the material. The uses of scored portfolios are limited to small formats and items that will not be handled often. Reinforcing the hinge with a fabric strip restricts its movement and cannot be recommended.

Uses

Materials

Heavy (up to 400 gm^2) cardboards can be treated as described on page 36. For portfolios you can use bristol board, museum board, cover board, or illustration board. For smaller folders you can also use Mi-Teintes paper or one-ply cardboard.

Steps

The steps in making a portfolio are as follows:
1. Choose the size.
2. Draw a pattern.
3. Cut (see pages 39–40).
4. Crease or score.
5. Fold (see page 36). The creased side faces outward.

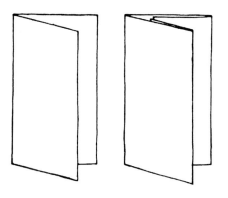

Two portfolios, creased at the back. One has front flaps.

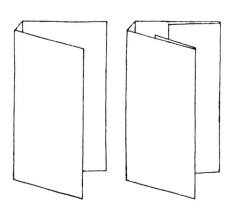

Two portfolios, creased or scored twice at the back. One has front flaps.

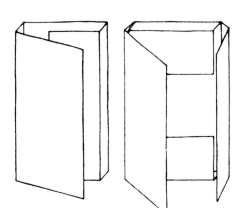

Two portfolios, one with a front flap, one with three flaps.

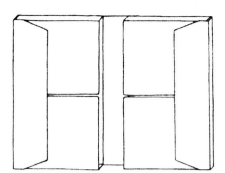

Double portfolio

Front flaps, unlike top and bottom flaps, should be slightly mitered to protect the corners from wear.

Cutting

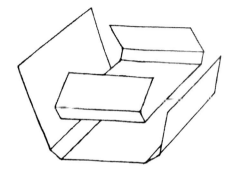

A portfolio with three flaps could theoretically be cut from a single piece of cardboard, but it would waste material and be rather complicated to handle. Cutting all three flaps separately is aesthetically unpleasing, and the papers inside the portfolio could easily get caught on all the resulting edges.

Gluing

It is best, therefore, to cut the top and bottom flaps as a continuous piece, the middle part of which serves as a reinforcement of the back cover. Connect the parts with dots of PVA, applied along the edges of the cover.

Details

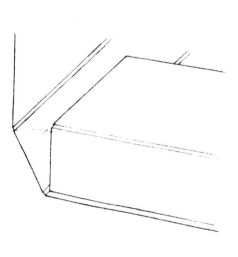

The portfolio in the drawing has a front flap that is a different width than the top and bottom flaps. This construction is necessary because the front flap has to be able to rest on the top and bottom flaps. The adjustment is unnecessary if the portfolio is made of very thin materials.

A PORTFOLIO WITH TIES

This portfolio could not be easier to make. Choose two rectangular pieces of cardboard, cut slits, and loop two ties through them. The width is easily adjustable.

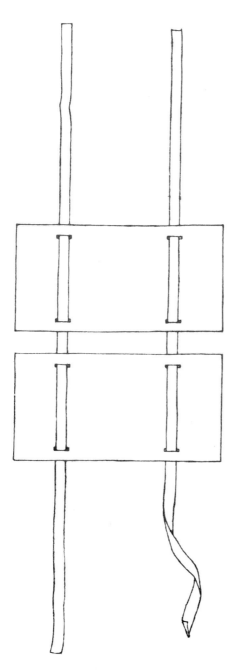

With a knife, cut the boards to at least the same size as the intended contents, sand the edges, and round the corners slightly. The boards can be covered and lined with paper or cloth (see page 53).

Almost any type of ribbon can be used for ties, as long as it seems strong enough. To calculate the length needed add: twice the width of the cover plus twice the maximum thickness of the finished portfolio plus twice 8 in (20 cm) to tie. Cut the ends at a slant or dab on some PVA to prevent fraying.

Cutting slits

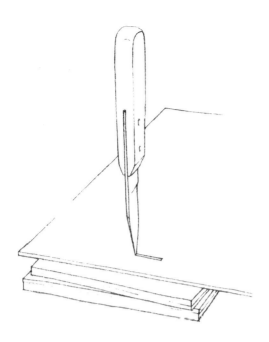

Use a chisel and hammer to cut slits the width of the ribbon. First mark the position of the ties on the cardboard. Each slit requires two parallel cuts about 1/8 in (2–3 mm) apart (see page 186) and two short cuts at the narrow ends, done with the tip of a knife. Lay several layers of cardboard under the portfolio and work on a solid surface.

A PORTFOLIO WITH A CLOTH SPINE

Measurements

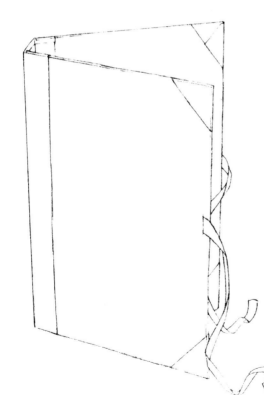

The desired measurements should be increased on three sides by 1/8 in (3 mm) to provide extra protection for the papers inside. Only very rarely will a cover be cut to the exact size of the contents.

The drawing on page 182 shows two stages of a portfolio with a flexible spine. Page 159 shows a rigid spine, which is constructed exactly like the back of a hinged box.

Back

Cut the boards so their corners are precise right angles, and be sure the grain direction matches.

Covers

Spine

Let us make a portfolio for drawings. It will need a flexible spine (b), made of a strip of medium-weight (about 140 gm^2) paper exactly the same length as the covers.

Choice of cloth

The choice of cloth for the spine (c) depends on the purpose of the portfolio. Cut the material with a ¾–1½ in (2–4 cm) overlap, which will connect the spine securely to the covers. The distance between covers and spine is one and a half times the cardboard thickness. With these measurements you can calculate the width of the strip. Its length should be that of the covers plus a ½ in (15 mm) overlap on top and bottom.

The inner strip (d) will be discussed later.

If the portfolio is large, attach ties and reinforce the corners now, before the parts are joined. Mark the positions of each element clearly with pencil lines on the inside of the cloth.

Marking

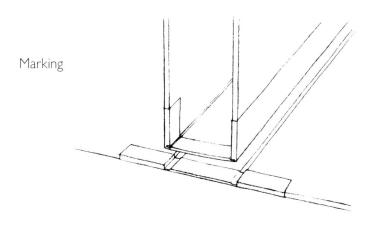

Gluing

Setting on

Rubbing

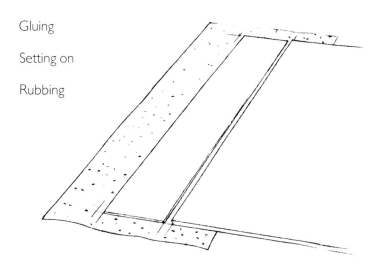

Now apply a 1:1 mixture of PVA and paste, place the parts on the marked positions, and check their alignment with a straightedge. Any corrections must be made before the adhesive dries. Press the three parts in position and rub with your palm.

Folding the overlap

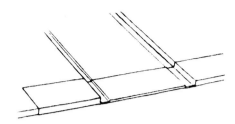

Position the parts so that the overlap sticks out over the edge of your work surface and fold the overlap over the edge of the spine (see page 56). Then rub the entire spine once more, protecting delicate materials from hands or the bone folder with a sheet of newsprint.

Inner strip

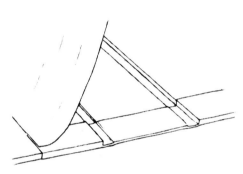

The inner strip (d) can be cut either from the same material as the outer strip, or from a lighter-weight material. The two strips are equal in width, but the length of the inner strip is about 1/8 in (2–3 mm) less than that of the spine. Remember that cloth will shrink in selvage direction.

Joints

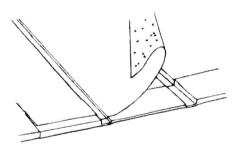

Using the same adhesive as before, set the strip down along its long edge, and work it into the joint. Keep the rest of the strip from contact with the spine until this step is completed, then proceed. Give special care to areas of contact between two layers of cloth. The spine will not last long if it is not made well.

Wide spines

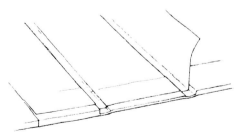

If the spine is wide, it is better to set the inner strip down the middle first and then work to the sides, into the joints, and to the covers.

Drying

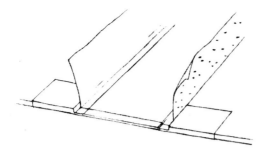

Both sides of the inner strip have to be held up while the work is in progress.

Spines, especially wide ones, need a minimum drying time of 2 hours and should be weighted between strips of cardboard.

CORNERS

Proportions

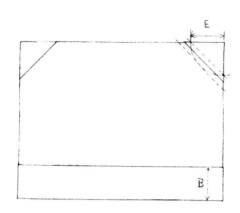

For a pleasing appearance, corner reinforcements should have some relationship to the visible part of the outside (spine) cloth strip. The drawing shows a pattern for a proportion I have tested and found good: the width B is the radius for a circle around the corner E. Draw the dotted lines at a 45-degree angle and find the edge of the corner reinforcement exactly between them.

Pattern

Gluing

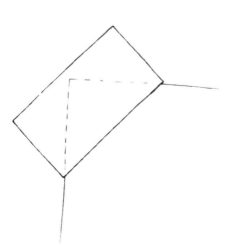

The corners are made from rectangular pieces of cloth that can be leftovers from the spine strips cut earlier. Grain direction is of no consequence here. Apply the same adhesive, set the cardboard corner into position as marked on the cloth, turn the cover over, and rub.

Pressing

Rubbing

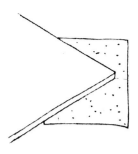

Turn the cover over again and, with the corner sticking out over the edge of the work surface, make the first fold over the shorter side of the cover (top or bottom).

First fold

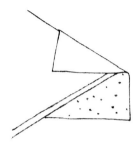

Pinching

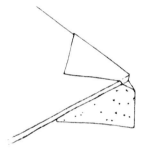

Never forget to handle the corner tips as described on page 59.

Second fold

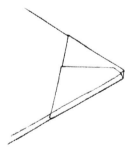

Now fold the second half of the corner triangle. Smooth the edge of the cloth with the bone folder and tap the corner gently. A somewhat rounded-off tip is less easily damaged.

TIES

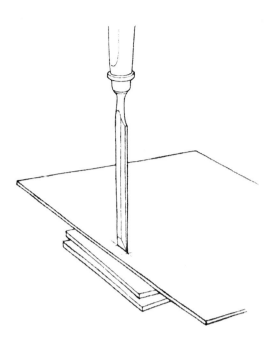

The weight and size of the portfolio determine the width, strength, and number of ties. The ties should be at least 8 in (20 cm) long, or longer, depending on the maximum thickness of the portfolio. The slits should be at least ½–¾ in (1–2 cm) from the edge. Cut them with a chisel and hammer, working on the right side of the cover over several layers of cardboard. The size of the chisel should match the width of the tie. For narrow ties, a single cut will suffice.

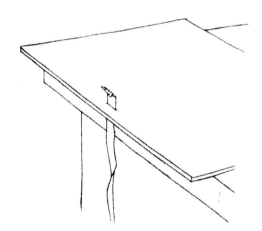

Push the ties through to the inside and glue ½ in (15 mm) of it down with emulsion. Two or three carefully placed blows with a hammer will flatten the edges of the tie and secure it in the slit.

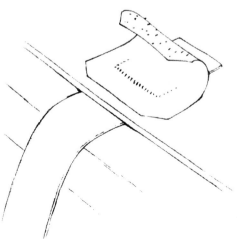

Finish by covering the end of the tie with a 1 × 1 in (3 × 3 cm) square of paper of the same color as the cover. It looks better, and papers will not get caught on this spot.

A PAPER-COVERED PORTFOLIO WITH A CLOTH SPINE

The basic construction of the portfolio is the same as before. The ties have to be installed before the covering.

Choose the paper and cloth carefully. Ties and lining should harmonize, and the whole should form an aesthetic unit whether through contrasts or harmony. This requirement is a very basic one and should always be met.

Variables include the width of the cloth spine strip, the proportion of areas covered by cloth and paper, and the materials. The paper may stretch across most of the cover or over as little as two-thirds of it.

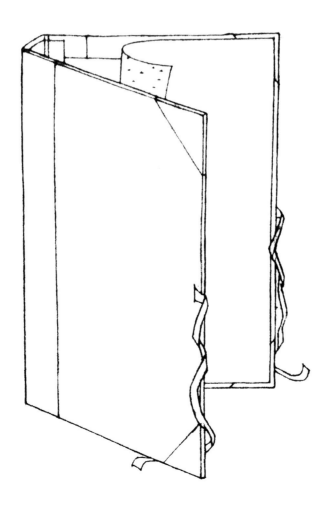

Inner strip

The inner strip overlaps about ½ in (15 mm) onto the covers. Most of it will be hidden by the lining.

Corners

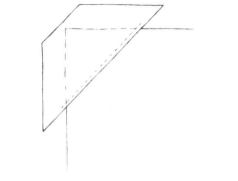

The pattern for the corner reinforcement differs from the one described previously. Here the shape is trapezoidal to match the edges with those of the cover paper. They fall onto the dotted line on the outside. For the size of the corner piece, see the instructions on page 184.

Pattern

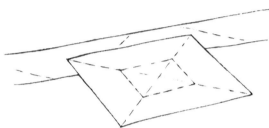

This pattern shows a method of cutting four corner reinforcements from rectangular or square scraps with as little waste as possible.

Setting on
Rubbing down

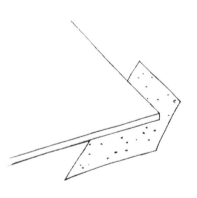

Set the cardboard onto the cloth, adjust its placement, and rub.

First fold

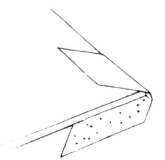

Fold around the top or bottom edge first.

Pinching

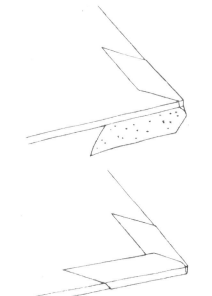

Pinch the cloth between the tip of your index finger and your thumbnail to fit it around the corner, as shown in the drawing.

Second fold

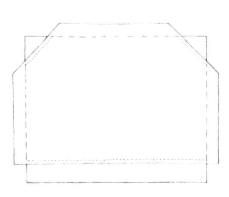

Make the second fold, rub with a bone folder over the cloth edges, and round off the corner slightly.

Covering

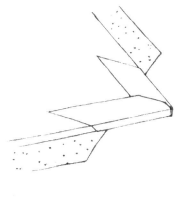

Cut the paper for the covering with a ½ in (12–15 mm) overlap on three sides and a slight overlap, about ⅛ in (2–3 mm) over the cloth strip.

With two short marks that will later be hidden under the paper, indicate the line along which the paper is to be set onto the strip.

Corners

Miter the corners at a 45-degree angle, and try to estimate how much the paper will contract. This amount will, of course, be different depending on the material used and the size of the piece.

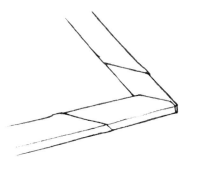

Gluing

Apply a 1:1 mixture of PVA and paste to the paper, set the cover on it, and rub. Then turn the portfolio over and fold the overlaps. Check for mistakes, turn the cover over again, and finish by rubbing the front. A detailed explanation of all these steps is to be found on pages 47–54.

Attaching ties

Every move should be planned, nothing left to chance. Give special attention to areas where paper is glued to cloth, that is, on the spine and around the corners. Clean newsprint should always be at hand for emergencies.

The widths of the overlaps match those of the cloth corners. When the paper covering material is in place, the ties can be attached (see page 186).

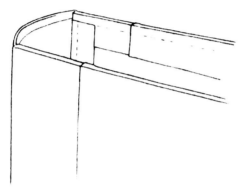

Lining paper

Gluing the lining

This is how the portfolio looks after covering and folding. The dotted lines indicate where the lining will end.

Upper and lower edges meet the edges of the inner cloth strip, and a border of the same width should be planned at the front edges of the covers. Again the probable shrinkage has to be taken into account.

Attach the lining with the same adhesive as the paper covering. Apply adhesive twice, and set it down along the front edge. Before you rub, make sure that the border is the same on all four sides. If that is not possible, a wider border is least disturbing at the front edge.

Drying and weighting

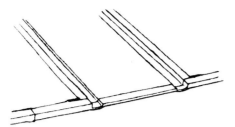

Keep the portfolio open for 12 hours or longer and weight it between cardboard blotters and wooden boards. Be sure the ties are not caught between layers during drying.

SMALL CORNER REINFORCEMENTS

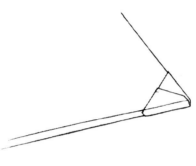

Nowadays almost invisibly small corners are often preferred over large ones. This is particularly true if the part of the cloth strip that can be seen on each side constitutes less than one-eighth of the area.

Invisible corners are of no consequence for the appearance of the portfolio, but they still serve as reinforcements. Cut them as rectangles and mount them as before. The covering paper, however, is trimmed not before but after it is attached to the board. In this case, do not press it down at first in the corner areas, so that it can be lifted off with ease for trimming. During these steps the portfolio faces inside up, and the corners stick out over the edge of the work surface to make them accessible.

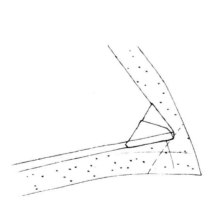

After trimming, press the covering paper in its final position at both corners and fold the overlap.

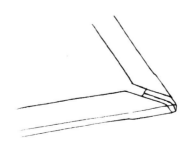

A FULL-CLOTH PORTFOLIO

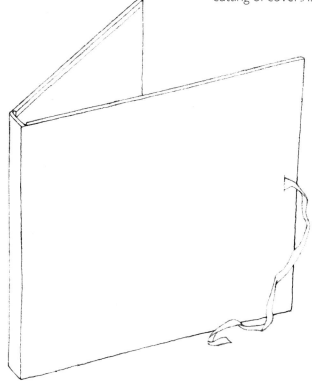

There are no new technical problems in this project, except the ones related to the handling of large pieces of cloth, which requires concentration. See pages 181–182 for cutting of covers and lining.

Cutting

Grain direction

Choose a piece of cloth large enough to include ½ in (15 mm) overlaps on all sides. Pay attention to the selvage direction and the fact that the material will shrink somewhat, which is especially significant in larger pieces.

Overlaps that are not sufficiently wide do not look good and are difficult to work with. If the cloth is unsized, see page 72.

Marking

On the inside of the cloth mark the position of the covers with lines that are dark enough to remain visible under the milky layer of adhesive.

Corners

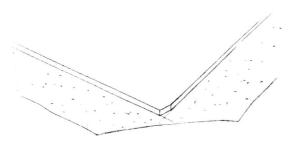

For heavy cloth, the corners of the cardboard should be mitered with a sharp knife, as shown. Sand the edges but do not round them.

Gluing

Use a 1:1 mixture of PVA and paste, apply it to the cloth twice with a round brush, and set the cover onto the marked areas. Turn the piece over and rub on the other side with your palms, starting in the middle as described on page 51. On this side do not push the material into the joints!

Some materials tear more easily than others, especially when they are damp. As a safety precaution, lay a protective sheet of paper over them before rubbing.

Another method

Another way, easier for beginners to follow, is to apply the adhesive to the boards instead of the cloth. Apply adhesive to each one and set it in place separately, then turn the whole piece over and rub on the cloth side. Finally, fold the overlaps (see page 56).

Rub all the parts one more time, trim the overlaps, if necessary, to make them equal, and proceed as directed on page 60.

PORTFOLIOS WITH FLAPS

Flaps protect the contents of a portfolio from dust, tears, light, and humidity. Each one of the portfolios described earlier can be constructed with one or more flaps. Following are instructions for attaching a side flap and for adding top and bottom flaps in one piece.

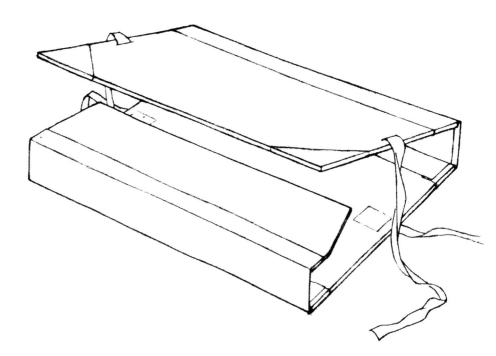

A PORTFOLIO WITH A SIDE FLAP

Choice of material

Measurements

Choose cardboard of the same thickness as the portfolio covers, or a somewhat lighter weight.

The flap is as long as the portfolio and it can be almost as wide as the entire portfolio, or as narrow as a quarter of that width. The top and bottom edges of all flaps but the largest ones are mitered, to prevent the corners from showing when the portfolio is closed.

The flap can be attached to the back cover in two different ways: It can be finished and then set onto the board (Method 1), or held in place by strips that hold the front and back of the board (Method 2), as illustrated in the two drawings on the facing page.

Gluing

Method 1

Finish the flap and glue it onto the board.

a. uncovered flap
b. back cover of the portfolio
c. spine
d. outer cloth strip
e. inner cloth strip

Method 2

The second method seems to be a more organic arrangement. An inner and outer cloth strip enclose the edge of the back cover. The resulting construction is similar to that of the spine of the portfolio.

No detailed description of an uncovered portfolio with flaps is given here, because besides the cover there is no difference between it and the covered portfolio.

Ties

Personal taste and the size of the portfolio are the criteria for the number and position of the ties. Some portfolios may be closed securely with a tie each on top and bottom; larger ones will need a tie on each side, whereas a portfolio with three flaps might not need any tie at all.

Hinge material

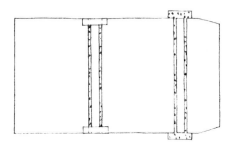

Use medium-weight (150 gm^2) paper for the hinges. If the flap is to be folded under the front cover, the width of the joint will be smaller than in the case of a flap that folds over the front cover. Another variable is the thickness of the cardboard.

Covering

Folding

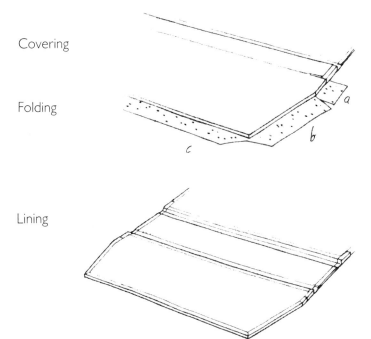

The flaps are attached to the back cover in exactly the same way as the spine (see page 182). Cover them in the same material as the portfolio; details are shown in the drawings. The covering paper overlaps the edges of the hinge strips about ⅛ in (3 mm). The side overlaps are ½ in (12–15 mm) wide. First fold a, then b, then c.

Lining

The lining, because of its irregular shape, has to be cut and checked carefully, since unequal borders would detract from the desired effect.

Cloth-covered portfolio

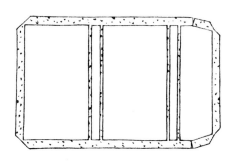

After you have made a two-part portfolio (pages 190–192), a full-cloth portfolio with a flap should pose no problems.

The drawing shows how the pieces are laid out on the cloth.

As before, ties can be attached after the flap is covered, but before it is lined.

PORTFOLIOS WITH THREE FLAPS

Proportions

Measurements

Rigid spine

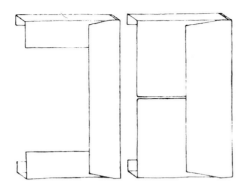

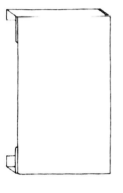

The proportions of the three flaps can vary quite a bit. Practical considerations are as important as aesthetic ones. Three possible solutions, among many, are shown in the drawings. For greater clarity the portfolios themselves are not shown.

The best way to arrive at correct measurements is to stack all the contents that should fit into the portfolio and measure all dimensions without applying any pressure on the stack. The height corresponds to the width of the hinges for top and bottom flaps. The hinge of the side flap is wider by the thickness of a cardboard layer. The width of the spine is the height of the contents plus two times the board thickness.

If the contents are not expected to change, construct a rigid spine, as in the hinged box.

Drawings of two sides show the basic construction. The important parts are the joints, where you can see, for instance, that the back hinge has to be higher than the contents by four times the thickness of the flaps and covers. The covering should also be taken into account.

It is important to notice that the hinge strip is not located between the front and back covers, but is flush with their side edges. The width of the hinge strips is reduced less than expected because of this arrangement (see the drawing on page 159).

Long-axis view

This view shows clearly that the top and bottom flaps are set in about 1/16 in (2–3 mm) from the edge of the bottom flap. The short

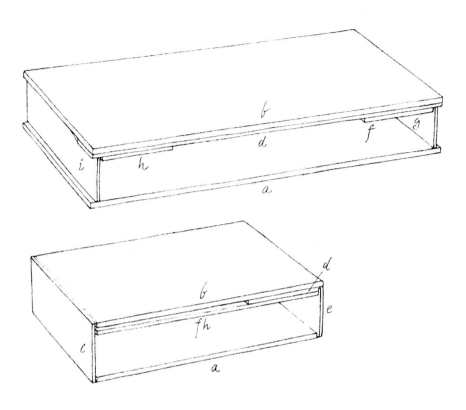

Short-axis view

a. back cover of portfolio
b. front cover of portfolio
c. spine or back
d. side flap
e. side flap hinge
f. top flap
g. top flap hinge
h. bottom flap
i. bottom flap hinge

axis view shows that the back cover of the portfolio is narrower than the front cover by a board thickness.

Make your calculations in the most practical way. Wherever possible, put the materials together and measure them. You can use tick marks on a strip of paper to transfer a measurement. The larger the portfolio, the less important such factors as the thickness of the material.

Material

The outer hinge strips can be cut from the same material as that used to cover the spine. You can cover the top and bottom hinges with the lining material of the spine, and line all the flaps the same way.

Outer strip

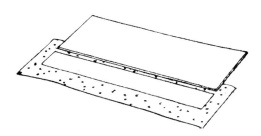

The strip that covers the flap hinge on the outside overlaps onto the flap ¾–1 ½ in (2–4 cm), depending on the size of the portfolio. On the other side at least ¾ in (2 cm) is needed to secure the flap to the back cover.

Folding

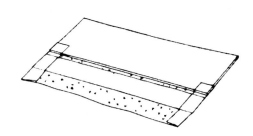

Cut the strip with ½ in (12–15 mm) overlaps on top and bottom and mark the placement of the cardboard pieces on the inner surface. Apply adhesive and position the cardboard. Fold the overlaps and work them tightly into the joints.

Inner strip

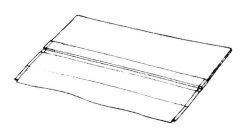

Cut the inner strip ⅛–¼ in (4–6 mm) shorter than the hinge, apply adhesive, set it in place, and rub.

Covering

Lining

Drying

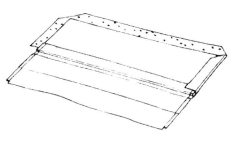

Covering material and lining are applied in the same way as the back and front covers. The overlaps are the same size. Weight the flaps and let them dry completely.

Hinges

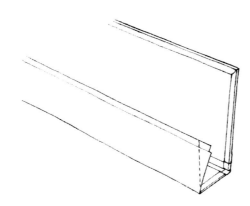

Hinges need special attention. Before you attach them to the covers, open and close them vigorously, especially the one near the back cover. The last step before you attach the hinged flaps is to miter the corners of the overlap (see dotted lines).

Ties

Now is the time to attach ties.

Attaching flaps

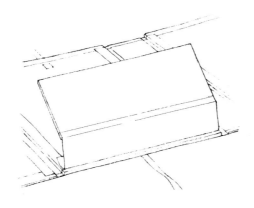

Use unthinned PVA and apply it generously, particularly where two layers of cloth will meet. To set the flap on, bring it to a folded position, flush with the back cover at both sides and slightly inward of the long edge. After a minute or so open the flap and rub the overlap by hand and with a bone folder. The cloth edge has to be rubbed down well, because it will remain visible through the lining.

Weighting

Drying

Weight the flaps in open position. After a drying time of 15 minutes, affix the lining.

The edges that are caused by the ends of the overlap under the lining can be very obtrusive. To avoid this, glue a piece of cardboard into the space on the back and front covers that is bordered by the overlaps. It has to be cut to exact fit and from board that is the same thickness as the cloth that forms the overlap. Use pure PVA as the adhesive. If you choose this option, do not attach the linings until everything is completely dry, and the amount of tension through the additional piece becomes apparent.

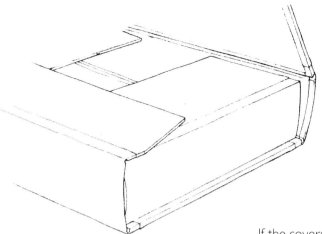

Lining

Portfolio with
three cloth-covered
flaps

If the covers remain fairly flat, use pure PVA, rather than a paste-containing mixture, as adhesive.

A portfolio with three flaps covered with cloth is constructed just like a portfolio with only a side flap (see page 194). After the ties are attached, the top and bottom flaps can be added.

BOOK FORMS

An interest in the history of book forms and in general with the history of what people have written on, sets the role of the codex, the book form that has become familiar during the last few hundred years, in a new light.

Earlier cultures were influenced by the surrounding landscape, the prevailing climate, and their mineral wealth. So were the tools and other utilitarian objects they produced. A thousand years ago symbols were carved in rocks, shells, and bark. Signs were written into the dust before a hut or a tent and disappeared again.

Strings and knots also served the needs of communication. The Mesopotamians imprinted their signs in clay; Egyptians wrote them onto long strips of papyrus. Asian Indians used prepared palm leaves — strings were looped through holes and the leaves tied into a "book." In China and Japan scrolls and accordion books were in use.

By no means is the Western codex to be understood as the apex of this development. Each separate form must be seen in its own context, with all its inherent advantages and disadvantages.

It is deplorable that many book forms have been brought to the brink of extinction throughout the last 150 years by industrialized book production. As in many other fields, a rich diversity has given way to sterile monotony.

In our time the means of communication have stretched far beyond the printed word — a development with unforeseeable consequences — but can this outweigh the loss of not only material forms, but also a way of life?

There are without a doubt a great many beautiful, even splendid, volumes bound in leather or parchment, gold-embossed, ornamented. They were made as late as the beginning of the industrial revolution.

These are the ideals that many a would-be bookbinder aspires to, and any exercise or simple project may seem a mere impediment along the way. More than just one error leads to this attitude. But it is difficult enough to find a justifiable purpose for a luxurious volume in today's world, and such an undertaking requires far more technical skill and knowledge than could be acquired during a short period of time. It is better to leave grand projects to the hands of masters. A modest design is not improved by a gold and leather treatment — the discrepancy only becomes more embarrassing for the knowledgeable observer.

Beauty is hidden in simple things, and can be realized with simple means. Precious materials can be used to mediocre ends, while convincing results can come from meager sources.

This is what we should aim for: few but well-matched materials, a feeling for appropriate uses, proportions, and color combinations, and the skill to handle the material correctly. Enough difficulties are presented by these requirements.

Historic book forms are introduced here not to be copied, but as a model for the design principles they represent. Some of these early forms have equivalents in modern ones: The scroll appears in electronic adding machines, the palm-leaf book as a wallpaper sampler, the accordion book as an advertising brochure, the loose-leaf book as a file folder.

Anyone who wants to find out more about historic book forms should consult the Bibliography.

THE SCROLL

In antiquity, scrolls made from parchment or papyrus were the preferred media of writing. The longest one that has come down to us measures 40 yd (40 m). The 8–12 in (20–30 cm) wide strips were rolled up with or without the aid of a stick or dowel. Occasionally, both ends were attached to such sticks. Only the part that was to be read was unrolled; the rest remained hidden. This system made it difficult to find a particular place in the text, and after each section the scroll had to be rewound.

Nevertheless, this book form remained with us for thousands of years. It was replaced by the codex in Europe during the fourth and fifth centuries, but many magnificent specimens are preserved. Whole library collections, such as the ones in Pergamon and Alexandria, consisted of scrolls.

In the Far East, silk or paper scrolls were in use from about 200 B.C.

The model for the drawing below is a Hebrew manuscript on paper with a cover that was probably added later, but complements the scroll perfectly. The illustration can give only a faint impression of the original.

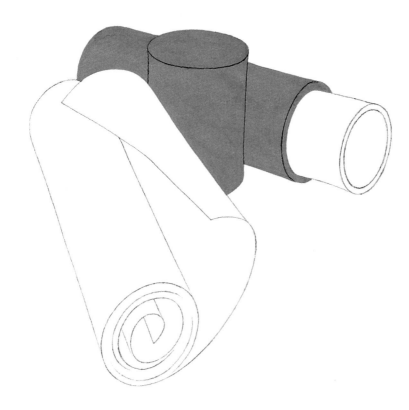

Material — Choose a medium-weight (100–140 gm²) paper, if possible from a roll, which would make seams unnecessary. If sheets have to be joined, the seams should be across the grain.

Cutting — The width of the scroll should be 4–12 in (10–30 cm). Cut the paper with scissors or a knife along pencil guidelines, and erase all lines afterwards.

Gluing — The length of the scroll depends on its purpose. Join two sections as described on page 167. Connections should be almost invisible, and practice on scrap paper is recommended. Use paste, and weight the piece while it is drying.

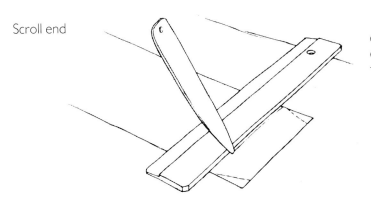

Scroll end — At the beginning of the scroll crease 1–2 in (3–5 cm) from the edge, miter the corners, and fold tightly. Then reopen the fold.

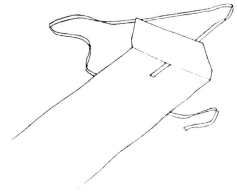

Tie — Guide a tie through a slit in the middle of the fold and fix it in place with PVA. Its length should be three times the circumference of the scroll when it is rolled up.

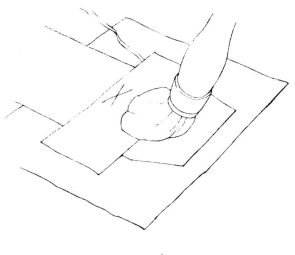

With a protective sheet in place at the fold, apply a thin layer of PVA to the end part only. Then fold, rub, and let dry under weight, to keep the end of the tie flat.

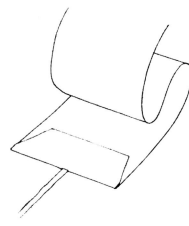

This is what the reinforced end of the scroll looks like. The other end can be treated the same way, but the fold must not interfere with rolling.

Cover

To store the scroll, roll it and secure it with the tie.

If a tube cover is planned, the tie is not necessary. Instructions for a cover can be found on page 172 — a round box with a lip. You must line the box before rolling it, because its interior will be inaccessible.

Reinforcements

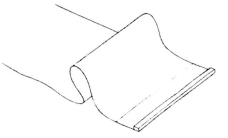

Both ends of the scroll can be reinforced with wooden sticks, as in Japanese and Chinese examples. They should never appear heavy.

THE PALM-LEAF BOOK

Palm-leaf books have been used in India for more than 2,000 years. They consist of a stack of palm leaves cut into long rectangles. The preparation of the leaves consists of drying, boiling, and sanding to make them usable and durable.

The delicate characters, almost exclusively in rounded shapes, are scratched onto both sides of the leaves. For better legibility, they are blackened with soot. The result is most astonishing.

The stack is often protected top and bottom with substantial covers of wood or, rarely, metal. A hole is drilled through one or two spots to connect everything.

The pliable leaves and the rigid covers form a perfect unit in spite of the startling contrast. Covers can be more or less richly decorated; the simpler ones are often more appealing.

Our intention here is to recreate the character of the palm-leaf book — the extreme contrast between the length and the width of its shape, and between the softness of its leaves and the stiffness of its covers. Our goal should be to create an aesthetic whole, with balanced materials and dimensions.

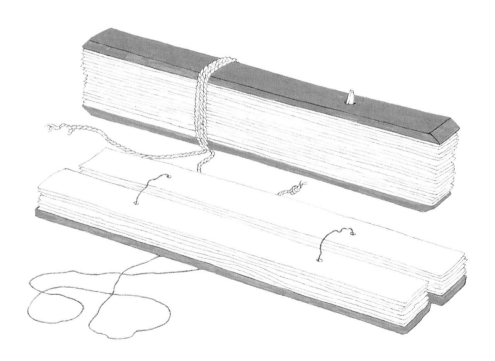

Material

Choose medium-weight (about 140 gm^2) material that can serve as a smooth writing surface. Drawing paper, maybe in beige or a yellowish gray, would be appropriate.

Assemble the covers from several layers of gray cardboard (80–120 gm^2) and pick a covering material in a color complementary to the paper leaves.

The tie should not be a bright and glossy ribbon; regular twine or cord is preferable.

Cutting the leaves

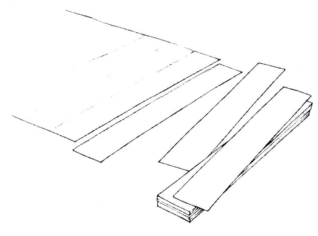

Each single leaf is cut with the grain and consists of two layers glued together with paste. This makes their texture close to that of palm leaves. Attach whole sheets of paper to each other and cut them into strips after they dry. The sheets should now be perfectly flat (see chapters about cutting, gluing, and drying at the beginning of the book). Two suggested formats are 2 × 12 in or 3 × 14 in (5 × 30 cm, 7 × 35 cm). Palm-leaf books are often rather voluminous and frequently considerably higher than wide.

Gluing the covers

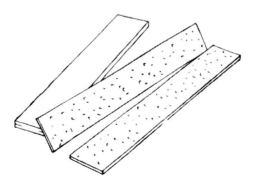

The ¼ in (4–6 mm) thick covers are assembled from several cardboard strips, cut with the grain and glued together like the sandpaper file described on page 66.

Sanding the edges

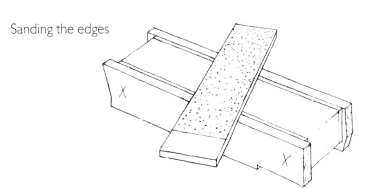

Place the leaves between the covers, protect the covers with two wooden boards, and insert the stack in a vise. Using a sandpaper file, and gradually finer grades of sandpaper, sand the sides until smooth. Do not round the corners!

Coloring the edges

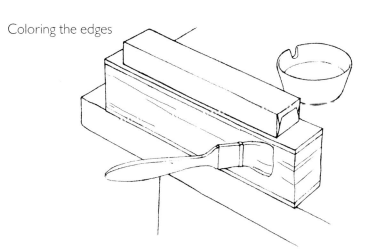

The edges of the leaves can be left plain or may be colored, as described on page 274.

Covering the covers

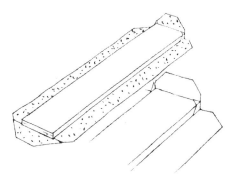

Covers that are more than 1/8 in (4 mm) thick should be covered as illustrated in the drawing. Drill the holes for the ties before covering to avoid damage to covers and linings.

Lining the covers

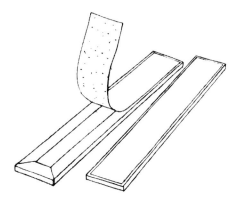

As lining, use either the cover material or the paper used in the book.

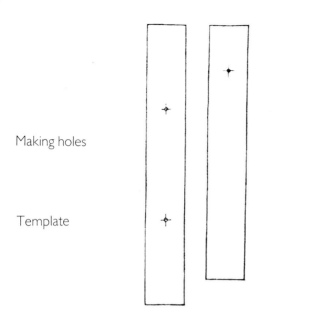

Making holes

Template

For the holes, use an awl or a drill and make them just large enough to accommodate the tie. To ensure that all holes are in the same position, make a cardboard template the size of the book and drill the first hole into it. Drill all following holes through this one.

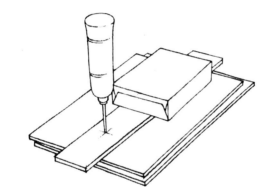

Binding

Working with awls, drills, and hammers requires a solid work surface protected by several layers of cardboard.

If there is only one hole, the book can be opened like a fan. In this case, fix the tie to the front cover and make a large enough knot at the other end to ensure that neither cover nor leaves come off. If the book is not in use, wrap the tie around it tightly several times. If more than one hole is drilled, several ties can be used, or a single one threaded through all holes. A handmade palm-leaf book deserves an equally individualistic closure.

THE ACCORDION BOOK

The countries of origin of this beautiful book form are probably China and Japan. Later it was adopted in Korea, Burma, Sumatra, and in regions of North and East Africa.

These books are made of paper or other material so as to form what is called an accordion fold. The covers are made from wood or cardboard, the latter always covered with paper or cloth. East Asian folded books always bear a label on the front cover. The covers may be flush with the leaves or overhang slightly.

The accordion book has many advantages over the scroll. Finding the right line in the text is less time consuming and, more important, these books require less storage space, a matter of consequence in a library like the one in Pergamon, where, during antiquity, up to 200,000 scrolls were stored at one time.

Our first project will be a flexible accordion book, folded from a long strip. Then we will make one with stiff covers and double layers of folds. Both projects are easy to complete, but a lot can be learned by dealing with materials and combining them in a new way.

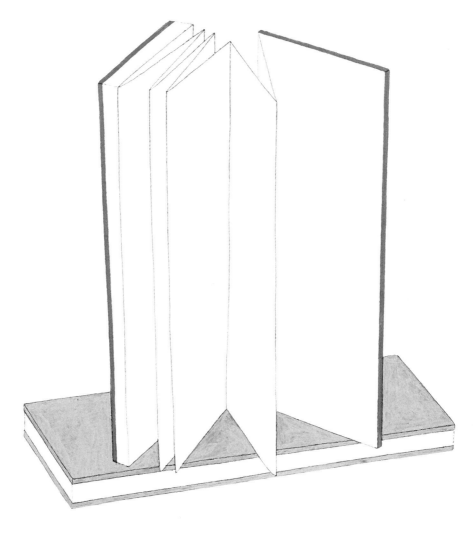

A FLEXIBLE ACCORDION BOOK

Format

First determine the format of the book. The height of the book will correspond to the width of the strip; on top and bottom add about ½ in (1 cm) for trimming. Cut two or three strips at the same time so that you can make the accordion longer during the work process. Cut the strips so that the folds will be parallel to the grain direction.

Grain direction

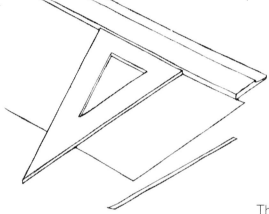

The first cut along the narrow side has to be made at exactly 90 degrees. A crease parallel to this cut marks the first fold. All following folds are made without measurements or crease. They turn out more exact this way.

Accordion fold

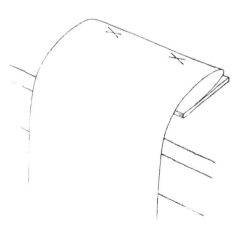

This drawing shows how the strip is easiest to fold: The unfolded part hangs over the edge of the table. Your palms and fingers lie symmetrically on the paper and press it slowly into the desired position. Look down onto the paper vertically and do not shift until finished. This unchanging view is the key ingredient for an even accordion fold; practice will do the rest. Trim off any incomplete fold at the end.

Connecting two accordion folds

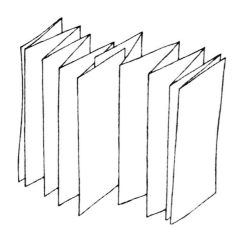

When two or more strips must be connected, the overlapping sheets have to be shortened just enough to keep the edges from the fold. Glue with a few dots of a water-free adhesive, because the connection should be invisible. The end sheets should face in the same direction, so it may be necessary to remove one fold at the end. (The illustration shows an extra fold, which must be cut off.)

Trimming

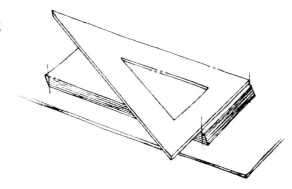

The last step is to trim the top and bottom edges. Guide a sharp knife along a straightedge or ruler as many times as necessary to cut through all layers, taking care to keep the guide in the same position.

Fashion a cover from medium- to heavyweight (160–200 gm^2) white or colored cardboard. Its two flaps will enclose the end pages of the folded strip. Top and bottom of the cover will be flush with the edge of the stack, or the cover could be about 1/16 in (1 mm) longer.

Cutting and creasing the cover

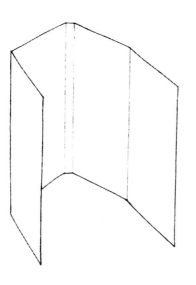

This accordion book, as opposed to the following one, can be opened only on one side. Take the measurement for the width of the back from the folded stack, squeezing it together a little. If the book is meant to hold photographs, the thickness has to be adjusted accordingly. Cut the cover slightly wider than the stack to accommodate the end folds in the flaps. Secure the flaps to the end folds with a few dots of adhesive applied to the inside of the fold.

AN ACCORDION BOOK WITH HARD COVERS

This accordion book is constructed with folded double sheets. The advantage is that you do not need long strips of paper and you can turn a series of easily available sheets of paper into a continuous accordion fold.

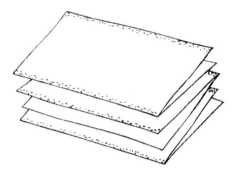

Origins

The drawing illustrates how the folded pages are attached to each other with hidden connections. Chinese and Japanese artists used this arrangement because their brush and ink drawings frequently bled through to the back side of the sheet, which made only one side usable.

Material

Raw and final format

Use soft, medium-weight (100–110 gm^2) paper to come close to the historic forms. The handmade papers of Chinese and Japanese models are unsized. Cut the sheets to the height of the book and twice the width, and add about 1/8 in (4 mm) for trimming. One extra sheet will be needed for endpapers.

Grain direction

Aligning

Weighting

Trimming

The grain direction is parallel to the fold. Crease and fold as described on pages 30–37. Align the folded papers on all sides, compress them with both hands, and afterwards between wooden boards. Weight the stack and let it rest for 24 hours.

The next step is to trim the edges. Use a template as described on page 242, which makes it possible to hand-cut all sheets with precision. Irregular sheets cannot be used further and have to be removed. The top and bottom edges are not trimmed until after the sheets are glued together.

Gluing

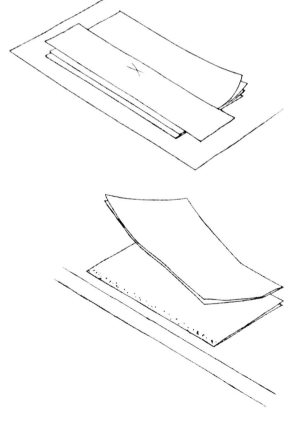

Take the first two sheets from the stack and arrange them on a sheet of newsprint to expose about ⅛ in (2 mm) of their folds. The technique is explained on page 73. Use pure paste and apply it evenly. If the paper buckles immediately after the adhesive is applied, it was probably cut in the wrong direction.

Remove the newsprint strip and place the first double leaf onto a clean surface, the fold facing you. Turn the second sheet so that the fold faces away from you and the surface with paste is downward. Set it flush onto the first sheet and rub both edges gently. If paste becomes visible along any edge, the application was too generous. Now rotate the pages 180 degrees, and set the third double sheet on, proceeding as in the previous step.

Drying

From here on the steps are repetitive until all the folds are lined up on top of each other in a perfect vertical alignment.

If the paper buckles more than expected, which tends to happen with very hard or thin papers, a blotter can be inserted between the layers, which are weighted and dried for an hour. Another reason could be too heavy or broad an application of adhesive.

Trimming

The previous chapter explains the process of cutting the top and bottom edges.

Cutting the covers

Covering the covers

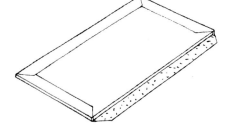

Cut the covers from cardboard. Make them the same size as the stack and cover them with cloth or paper in the correct grain direction (see pages 47–60). In the drawing only one edge remains to be folded, and the corners are already done.

It is not advisable to attach the cover board to the entire endsheet, since warping would be inevitable. Instead cover the sheet with newsprint, expose only a ½ in (1 cm) strip, and apply PVA. Then set the cover in place and press it on. The part not glued can be cut off after drying. These steps are explained in detail on page 282. Weight the book and let it rest for 5 minutes.

Attaching the covers

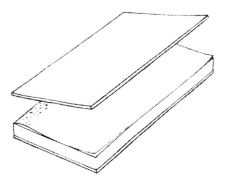

The linings should be of the same size as the folded sheets and should never stick out from under the cover board. They have to be cut small enough to allow for some expansion under the influence of the adhesive.

Lining

Since the linings are cut with the grain, their expansion will occur mainly in width. Use the same adhesive as for the cover, maybe adding some paste. Rub, weight, and let them dry.

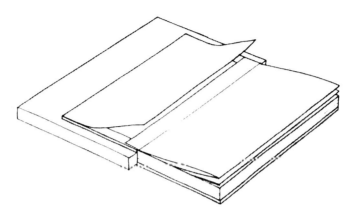

Inserting blotters

Drying

To keep moisture from entering the book-block, insert underneath each cover a cardboard blotter of somewhat larger size than the book. It should be pushed close to the fold. Put the accordion book between clean cardboard and wooden boards, weight it, and let it dry. After a day it is ready to use.

SIDE-SEWN BOOKS

Not much is known about the origins of this book form. Its earliest examples come from China and Japan between the tenth and seventeenth centuries.

Substantial literary works are always subdivided into smaller units such as the ones shown in the illustration. The parts are kept together in matching covers. There may be two wood covers tied together with ribbons, or the cover may be in three sections and closed with ties or other toggles. The subdivisions were necessary because the entire work would be unwieldy in one volume.

The books consisted of doubled-over sheets with openings at the back, so that each one of the leaves consisted of two outer writing surfaces and two inner sides that could not be seen after the leaves were sewn together. As explained before, Chinese and Japanese papers were ink permeable and could only be used on one side.

This book form is extremely simple and forbids any splendor. It appears in its most noble form through the use of balanced proportions and colors, cover surfaces, and titling.

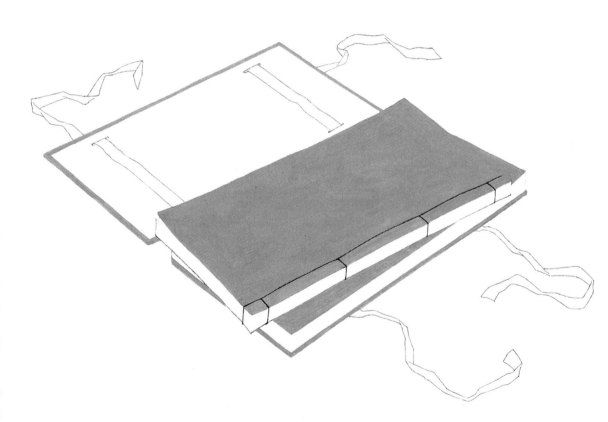

Side-sewing

In side-sewing the threads are sewn through and around the back of the book, as opposed to stitches that penetrate the folds of each layer. Side-sewing is the technique of choice when the sheets are not folded at the back or when you are dealing with single leaves. Besides thread, such materials as metal clips, spirals, rivets, and others can be used. None of these alternatives will be discussed here.

Choice of paper

For side-sewing you need thin and flexible paper that can still be opened after it is bound together. Ordinary typewriter paper can be used. The cover leaves can be double sheets of any desired material, including wrapping paper or cloth. The spine remains uncovered.

Covers

Another method is to cut two pieces of heavy (200 gm^2) cardboard to the size of the book-block, cover and line them, and treat them like the folded sheets. In Japanese examples covers and linings are usually stretched onto the boards (see page 68).

Thread

Sew with bookbinder's thread, thin cord available in many different colors, or buttonhole thread. Do not use synthetic materials. The only other equipment needed is a needle with an eye large enough to accommodate the thread.

Stitches

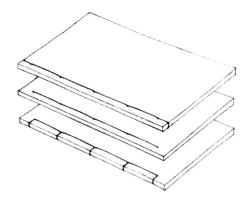

The number of holes needed is determined by the thickness of the book. The drawing shows five, done in three different ways. Other variations are possible.

Placement of stitches

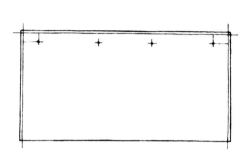

For a finished format of 7 × 10 in (26 × 13.5 cm), cut the paper with the grain and add 1/8 in (3 mm) on each side so that later the pages can be trimmed. The stitches should be set inward 1/16 in (1 mm) from the back, the first and last ones about 1 in (3 cm) from the top and bottom edges.

Folding

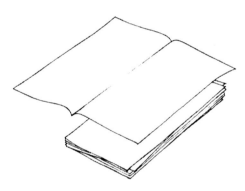

Fold as described on page 30. Repetitive movements should always be done in a rhythmical but concentrated way. As the last step, fold the two endpapers in the format of the book-block. Remember to cut them with the grain.

Aligning

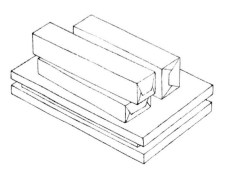

Aligning the stack can be a difficult endeavor if the paper is very lightweight. In this case divide it into several smaller stacks, align them, and reassemble. Squeeze folds together, mark the position of the stitches on a piece of cardboard the same size as the paper, and weight everything between two boards for a day. Put the weights on carefully: the layers should not be shifted.

Weighting

Wrapper

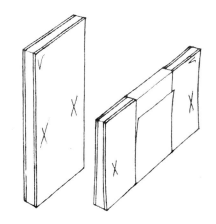

Decide which part is the top and write an identifying mark in the upper left corner of the front (V is used in the drawing). Then wrap a strip of paper around the book-block and align the pages once more.

Preliminary gluing

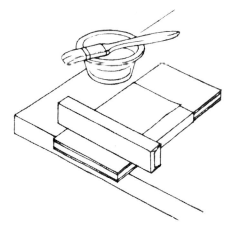

Apply pure PVA to the top and bottom edges of the strip, and remove the weights only after it is dry. This wrapper is temporary.

Trimming the back

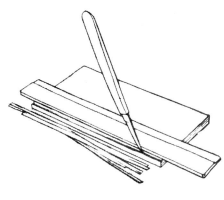

Since the book-block cannot shift any more, you now can trim it with a knife along a line marked on the back. Depending on the height of the stack and the kind of paper used, ten to twenty cuts might be necessary. The position of the knife should remain constant.

Gluing the back

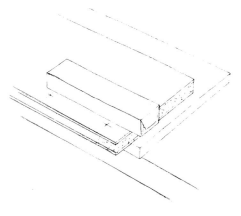

Position the book-block so it sticks out over the edge of the table, and weight it. Apply PVA to the ends of the spine, only as far as the outermost holes, which were marked previously. After the first layer of adhesive is dry, apply another one.

Trimming top and bottom

After ½ hour trim the top and bottom of the book-block so that it is held together only by the adhesive just applied to the back. Japanese bookbinders handle this temporary binding with wads of paper inserted in specially drilled holes and hammered flat at the top, but this method would be hard for us to duplicate.

Punching holes

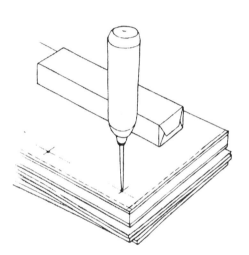

Over a thick layer of cardboard, punch the holes with an awl. They have to be wide enough to accommodate three times a double thickness of thread. During this step the book-block has to be kept from shifting. A paper strip wrapped lengthwise around it can be helpful.

Reinforcing the back corners

Corner reinforcements can be made from lightweight paper, either in the color of the book-block or matched to the cover paper. Apply PVA to the reinforcements and set them on with great care; otherwise such detail work becomes meaningless.

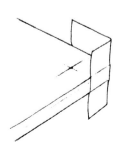 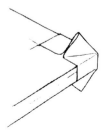 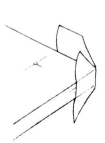 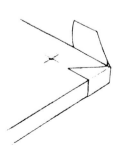 

Setting on cover pages

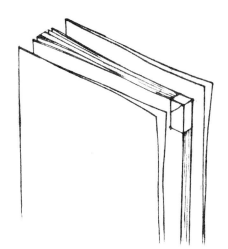

When the cover pages are cut, determine the position of the holes and punch them. Position the sheets with the fold in front. Before sewing, glue the cover pages lightly to the book-block near the back edges.

Sewing

The drawing shows the path that the thread takes through the back of the book. The thread can be used in single or double thickness; size the holes accordingly. The thread should never have to be forced through the paper.

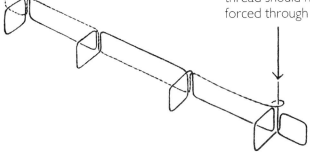

Keep the thread taut at all times. Tie the end immediately after the first stitch and hide the cut-off end between the pages. No decorative tassels are necessary.

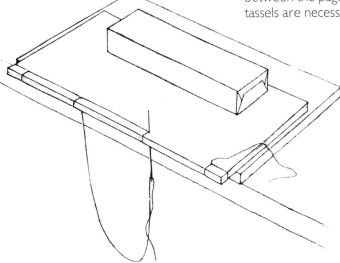

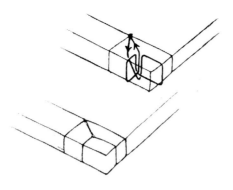

Instead of the corner treatment just described, a different and special treatment can be used. It is an elaborated version. Side-sewing and reinforcement can be done in one step and with the same thread. The drawing below shows three variations that should be easy to understand.

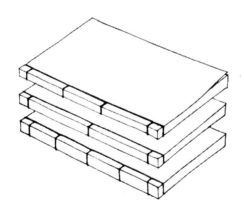

THE LOOSE-LEAF BOOK

Loose-leaf books were more than just a substitute for "real" books; they were and still are book forms in their own right. The loose leaves can be kept in place by covers of wood or leather and held together by string or leather straps. Some loose-leaf books are kept in a matching folder or a leather bag. We have examples of superb workmanship and beauty, gold-decorated and embossed. Others, plain and unadorned, are every bit as impressive.

Most loose-leaf books were used for religious purposes. They were prayer books or magic manuals, and in regions of East and North Africa or Tibet some of them may still serve their original purpose.

The loose-leaf book can be considered a predecessor of the codex, which is by no means a devaluation of the loose-leaf. Our familiar book is no more than a stack of leaves between two covers, maybe in a folder, except that everything is connected: the leaves to each other and to the covers. The loose-leaf book is actually closely related to the temporary binder and other file folders.

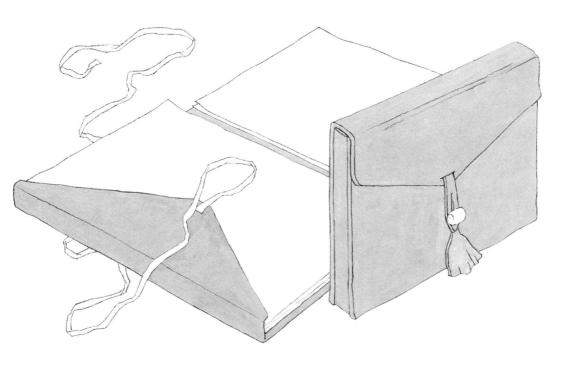

Uses

Paper

Today you can use loose-leaf books for recipes, sketches, a diary, or a collection. The quality of the pages and their format depend on the kind of project you are planning. You also have to decide whether the leaves should be single or double.

The chapter on portfolios explains how to create a three-part folder. Given here is just a description of how to make an expandable bottom fold. If you are thinking of binding the pages at a later time, folded sheets are preferable to single ones. Single leaves can be bound, but the preparations are more time-consuming.

Binding considerations

Ingres paper, folded three times, gives a very pleasing format.

Materials

You need cardboard, covering and lining paper, and cloth for the expandable bottom fold. All materials should be coordinated.

Measurements

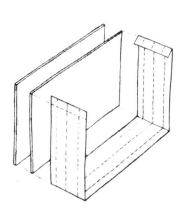

Lining the sides

Take measurements from the contents. For this project, assume it will be a 1 in (3 cm) stack of 10 × 6 in (24 × 16 cm) sheets. Cut the cardboard ½ in (1 cm) larger on each side and attach the lining paper, cutting the board to its final dimension after it is dry. As adhesive use a 1:1 mixture of PVA and paste unless otherwise specified in the instructions.

Cutting the flap

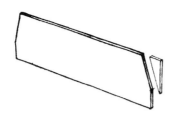

Cut the piece of cardboard for the flap. Its length corresponds to the width of the portfolio; its width is one-third of that amount. Miter the corners and keep in mind that the size and proportion of the corners will determine the character of the project.

Covering the cloth strip

The correlation between the cloth strip and the sides is explained in the drawing. On each side add ¼–½ in (5–10 mm) for overlaps and the same amount again for trimming after the strip is lined.

Drawing the pattern

Creasing

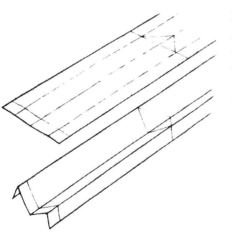

Cut cloth and paper with the grain, apply a thin layer of adhesive to both of them, attach them to each other, and rub. Let them dry for 4 hours between blotters and flat wooden boards. After the lined strip is completely dry, cut it to its final dimensions and mark the lines along which it will be folded: two long ones are marked on the outside, four cross folds on the inside, and the slanted folds on the outside.

Folding

Fold the strip according to the dotted lines in the drawing by hand or with the help of a bone folder, and rub the folds until they hold their shape.

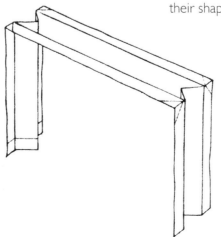

Connecting the strip to the sides

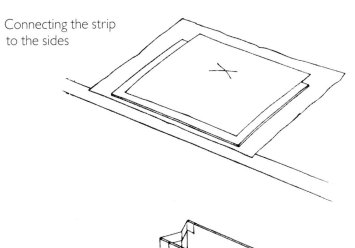

You need a rectangular piece of newsprint that leaves the width of the overlap free on the cardboard. With the lined side face down, apply pure PVA to the free edges.

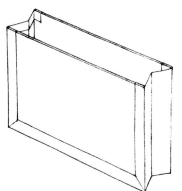

Connect one piece of cardboard at a time to the folded strip. It may be necessary to apply fresh adhesive to the strip. Treat the corners as described on page 116.

Covering the flap

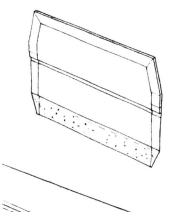

The basic construction of the flap is explained on page 196. Finish the flap except for the inner cloth strip and the liner. Both parts will be attached after the flap has been connected to the portfolio.

Attaching the flap

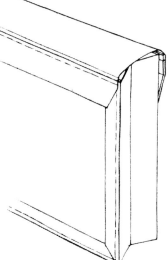

This view of the back of the portfolio shows how the cover flap with its mitered corners is attached to the back. This is done with pure PVA.

Reinforcing the strip with cloth

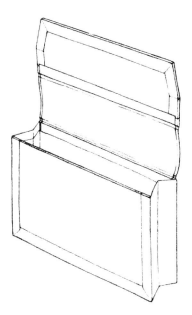

When the portfolio is open, the cloth strip that covers lining and hinges becomes visible. Its overlaps on the inside of the portfolio and on the flap are the same width.

Covering the outside

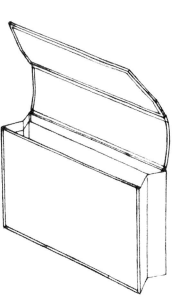

Drying

The last thing to do is to cover the outside of the boards. Leave about 1/8 in (2–3 mm) of the edge reinforcements visible on the sides and on the bottom; on the top edge fold over into the portfolio. On the back side the cover ends 1/8 in (2 mm) below the upper edge of the board, which is hidden by a cloth reinforcement strip.

The best way to dry such a portfolio is to fill it with a stack of papers between two sheets of cardboard, which prevent moisture from getting into the stack. Then dry the entire portfolio between blotters and boards for a day.

THE CODEX

"A stack of papers, held together on one side" — this is a somewhat simplistic definition of the codex; it would even include paperbacks, those poor relations of the book family.

Early in its development the codex was rather unimportant compared to the scroll, and we have to imagine it simply as folded papyrus leaves without cover, possibly with a reinforcement along the spine. The brittle papyrus was later replaced by parchment, which was easier to fold. Several layers of folded signatures could be connected with chain stitches, wooden covers were added for protection, and the book-block binding that we consider typical for the Western book first appeared after the Middle Ages. Later the spine was rounded, the whole book-block was cut instead of individual signatures, parchment and leather covers were ornamented, tooled, and gilded. This fascinating process can be traced from the third century A.D. to the present time, in which the codex has to defend its position against radio, television, tape, and microfilm, not to mention the computer.

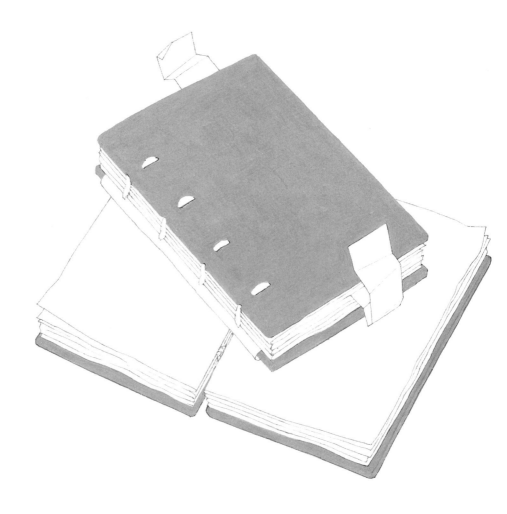

The codex is by no means the sole invention of the occidental world. The European examples were preceded by volumes of highest quality in Persia and Turkey, regions that continue to provide a valuable source of stimulation for this art.

The drawing depicts a Coptic manuscript on parchment from the eighteenth century. Its pages are sewn with a chain stitch and protected by rough-hewn boards. The cords are laced through the wood and provide the only connection between the book-block and the cover. The spine is neither glued nor reinforced. Of all preserved early books these Coptic bindings are the closest examples of the early European book forms.

PAMPHLETS

Even a simple pamphlet can be functional and good looking at the same time. Attitude and taste are essential, but both can be developed through practice on unpretentious projects.

The school notebook that everyone remembers is a questionable guide, but examples from the beginning of the early nineteenth century show all the hallmarks of good workmanship: covers in pleasing colors, such as grays with tints of red, blue, or yellow, labels placed deliberately, and other details, not modern at all, but worthy of setting a standard to strive for.

Unless you choose to color your cover paper for a particular requirement, the sometimes limited assortment offered by art-supply dealers will have to suffice. Finding the right materials can be a time-consuming process, and the perfect discovery is often made in the most unlikely place.

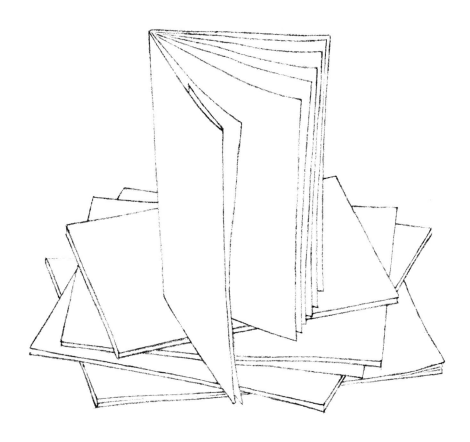

The pages of a pamphlet can be held together by adhesive or staples, but thread is usually the most appealing material. Notebooks consist, as a rule, of one signature with a cover. A signature, in turn, consists of several single folded sheets . . . .

Folded sheets

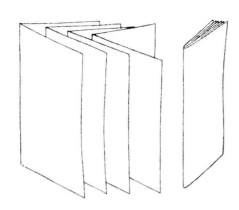

Folded and cut sheets

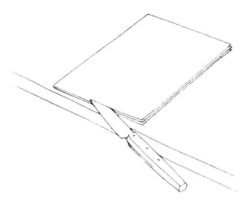

. . . or of a sheet folded more than once and cut open after folding (see pages 30–35).

Sewing

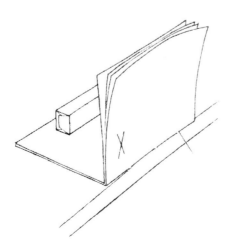

The first project described here is to sew together a signature of four to eight folded leaves and a cover, consisting of another folded sheet. Open the signature in the middle, align the fold with the edge of the table, and weight the booklet as shown. Holding the upright pages with one hand, punch three holes from the inside to the outside at an angle of 45 degrees. The distance between the holes is shown on the following pages.

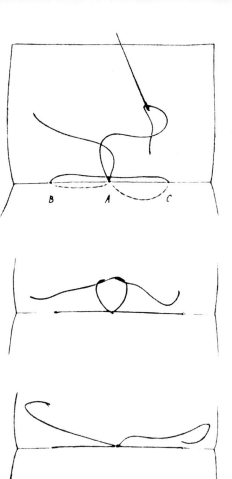

Thread the needle with a thread long enough to tie the ends when the sewing is finished. Tie half a square knot, pull the ends tightly, being careful not to rip the paper, and finish the square knot.

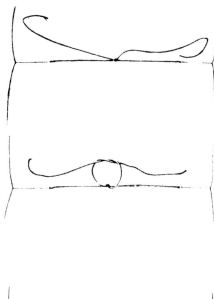

Tighten again; this time be careful that the first half of the knot stays in place. Now pluck the string gently to make sure that the tension is adequate. If it is too loose, the process has to be repeated. Otherwise, cut the ends about ¼ in (4 mm) from the knot.

The last step is to cut the front edge of the booklet at an exact angle (see text on cutting).

Cover with flaps

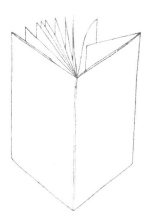

After the signatures are sewn together the cover can be attached. For this technique it has to be cut with two attached flaps, which will be folded around the two outer pages, which may be made of paper that is slightly sturdier than the other pages. The knot can be placed on the outside if it seems disturbing in the middle of the booklet.

Sewing two signatures together

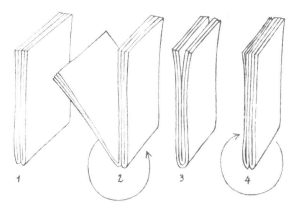

Two signatures can be sewn together in a very similar way (1). Fold half of one layer around the other (2) to turn two into one (3). After they have been sewn according to the instructions given earlier, fold the first signature back to its original position (4). Align the edges. Attach a cover with flaps, after creasing the spine with two parallel lines according to the thickness of the stack.

ADHESIVE-BOUND BOOKS

Most books can be categorized as either sewn or adhesive-bound. The latter method makes it possible to connect single sheets through the use of special adhesives. Books have been bound in this way since the nineteenth century, but it was the discovery of synthetic adhesives of high elasticity that made this technique widely accepted. Today there are numerous examples: paperbacks, address books, brochures, and the like, and even hardcover books are often adhesive-bound.

Advantages

The advantage of this binding is that single sheets can be bound without further preparation, whereas sewing requires folded sheets. Moreover, needles, thread, and sewing frames are unnecessary.

Disadvantages

Books that are adhesive-bound by a machine frequently cannot be opened with ease. The reason may be that the paper is too stiff for the format or it was cut in the wrong grain direction. Cheap adhesives can lose their elasticity, and pages begin to fall out.

Just as the weakest link determines the strength of a chain, so the durability of the spine depends on the connection between single leaves. The connection between book-block and covers is a problem that can never be solved completely satisfactorily in an adhesive-bound book, which may withstand the demands of the bookshelf but not those of the reader, and often does not satisfy our desire for adequately made objects. Adhesive-bound books should always be combined with covers of paper or very thin and flexible cardboard, which have to be set back from the book-block to make sure that they do not pull at it when the book is opened.

It is pitiful indeed to see how many a sewn book in need of repair has its spine chopped off and is turned into an adhesive-bound book, its wounds hidden under a cover, all in the name of saving time or money. Typescripts, however, or sketches torn from a pad, can be adhesive-bound to great advantage.

The durability of adhesive-bound books depends on the choice of adhesive and paper and on whether it was cut in correct grain direction. (Commercially produced books are not included in this discussion. Other factors are at work there and one small variable can affect thousands or ten-thousands of books.)

Unless a special formula for adhesive bindings is available, any good, pure PVA will suffice. The paper used should not be coated, but should have as little sizing as possible and a matt surface.

Two methods

Double leaves (folios) always produce a more durable and flexible spine than single ones. If you have a choice, it should be made in favor of folded sheets. If single sheets have to be joined, use one of two methods: If the spine is compressed during the gluing process, the adhesive cannot penetrate. The book will be easier to open, but durability suffers. If the spine is being manipulated and moved during gluing, the adhesive will seep deeper between the sheets and form a stronger bond, but the finished book cannot be opened completely. Pages cut in the wrong grain direction will aggravate this problem.

Trimming

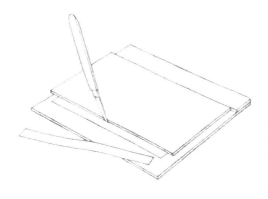

Unless the pages are already in the exact same size, as for instance typewriter paper would be, they have to be trimmed with the help of a cardboard template. Thin booklets can be cut after they are bound. It is important to cut exactly to make perfect alignment possible later.

Method 1

Aligning

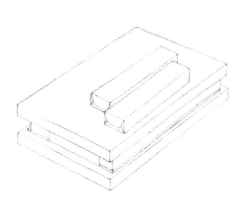

Pages that carry text have to brought into the right sequence, then aligned between two sheets of cardboard of the correct size, and held in position with a strip of paper wrapped around the entire stack.

Trimming the back

If any page does not reach the spine, even by a very small margin, the adhesive will not hold it in place and it will fall out of the finished book. The alignment of the paper is therefore a very important step.

The cut along the back has to be straight but not smooth. It should be made even rougher with the help of sandpaper, to make the surface more receptive to the adhesive. Remove any loose particles. During cutting and sanding the spine has to be compressed.

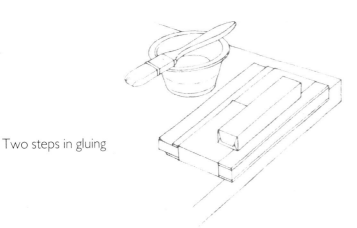

Two steps in gluing

Weight the stack, let the back stick out ½ in (1 cm) from the edge of the table, and check to see that nothing has shifted. Apply PVA to the spine in two layers, the second one after the first one has completely dried. All pages and the edges of the cardboard covers must be coated evenly. The second application can be thicker than the first one, but do not let the adhesive drip. The purpose of the first, thin layer of adhesive is to keep the second one from penetrating into the book-block.

Method 2

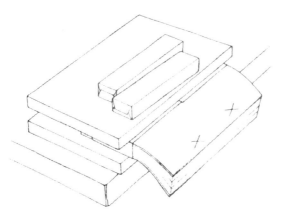

For the second method, the book has to stick out so far that you can easily bend it up and down while it is held between the boards. Apply the two layers of adhesive while the book is bent as far as possible in either direction and put the book back between the boards until the adhesive dries. The gluing can be repeated while the spine is immobilized.

Removing the boards

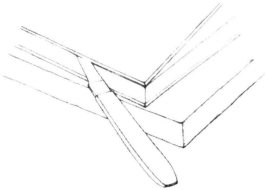

After the adhesive is dry, take the book-block from between the boards and remove the paper strip. Hold the book-block, open one cover to an angle of about 120 degrees, reclose it to an angle of 20 degrees, and separate it from the book-block with a sharp knife held exactly parallel to the book-block. Repeat this process on the other side without injuring any of the paper sheets.

Further steps follow the sequence on page 248.

Endpapers

Reinforcing the spine

Cover

White or colored endpapers can be attached to the book-block and will strengthen the attachment to the covers (see page 258 for an explanation of how to attach endpapers to the back edges). A strip of Japanese paper or thin cloth can be glued around the spine to improve its stability. For this procedure the adhesive is applied to the spine, the strip is set on parallel and symmetrically and rubbed on.

Now cut and crease the cover. The material must be thin enough to bend with ease or it will not adhere to the spine.
(see page 248).

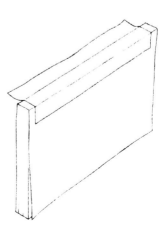

THE BROCHURE

The brochure is a book with a cover of a flexible cardboard or paper that is glued directly onto the spine. Brochures are flexible, and their binding is usually of a simple kind.

Thread, adhesive, or staples can hold the book together. Its edges can be cut or raw, and the cover can be creased two or four times, folded in front, on all sides, or not at all. Usually there are no endpapers, and the first and last signatures are not reinforced.

Financial considerations are usually responsible for the choice of a softcover, but in some cases it is the improvised appearance that is sought by using this method. The book with a rounded spine and two hard covers represents durability and demands a certain respect by its mere presence. Softcovers seem to imply actuality, practicality, and action. These observations are of course superficial; never forget that there are good and bad examples of both soft- and hardcover volumes.

PREPARATIONS

Grain direction

Use about twenty sheets about 18 × 24 in (46 × 60 cm) and fold each three times (see page 34). The grain direction should be parallel to the folds. If this is not feasible, cut sheets in half the indicated size and fold only once. Folds against the grain should be avoided wherever possible, or pages will not open easily and creases and ripples will appear.

Aligning

Mark the upper left corner of the front page to ensure that the book will always be positioned correctly. Align the edges while holding the stack loosely, allowing air between the pages. Thrust the stack sharply against a plane surface several times; no leaf should stick out on any edge. Compress the back, align it once more, and place the stack between two slightly larger boards.

Storing between boards

If work has to be interrupted for any length of time, always store the book between boards. The book is easier to manipulate, the alignment stays intact, and the boards provide protection from accidental spills in the work area.

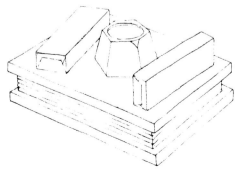

Weighting

The stack should remain weighted for several days: modest pressure for a long period is preferable to high pressure for a short time. The result is a compact stack of papers that already gives a foretaste of the finished book.

TRIMMING

Trimming the front edge

An untrimmed brochure will be hard to open. However, if trimming is desired, it must be done before the signatures are sewn together, as was the procedure for centuries before the advent of heavy cutting equipment. If cutting equipment is available, use it after the spine is glued. Use the following procedure: Cut a cardboard template somewhat larger than the desired format of the book, and a second strip, as shown in the drawing.

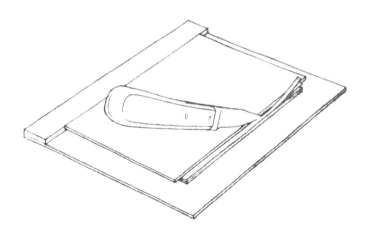

Trimming top
and bottom

One layer after another can be cut to size with the help of these templates. The cut signatures are aligned with each other and checked; they should be equal in size. To trim top and bottom, arrange the work according to the

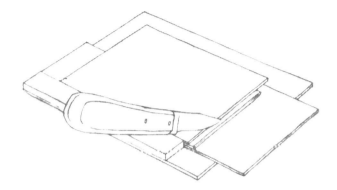

drawing. A movable underlay of cardboard will keep the knife from going over the same groove during each cut.

COLLATING

Signature marks

Before the signatures are sewn together, they have to be arranged in the right sequence. Needless to say, this job has to be done with care, since errors are irreversible. The sequence can be checked by looking at page numbers, signature marks, or registration marks.

Wherever signatures have to be taken apart for repair work or reinforcement, all pages should be checked for correct sequence before reassembly.

To check sheet numbers look at the signature marks, small digits in the bottom margin below the text at the left or right side of the first or third page of each signature.

Another method of marking the order of signatures is the placing of little marks or numbers on the outside fold of each signature, forming a stepped pattern.

SEWING

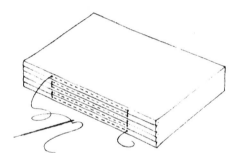

The simplest technique may be used. The drawing at the left shows the pattern; the first stitch is the one on the bottom right. Mark the length of the spine and the location of the stitches on a cardboard strip and transfer the markings onto the back of the stack with the help of a triangle, as shown below. The drawing on page 266 shows a good setup for your work table.

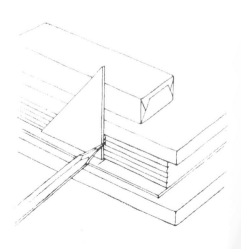

Thread

The strength of the thread that should be used is variable. The thread's resilience is less important than its thickness — there should be no visible difference in bulk between the edges of the book-block and the spine. Three factors have to be considered: the folds, which can be compressed by pressing; the bulk of the thread, which should be as thin as is viable; and the composition of the paper, which will absorb more of the thread thickness the softer it is.

Swell

The thickness of the signatures must also be considered. A book with its pages divided into twenty signatures will be thicker than one with the same number of pages arranged in thirty signatures; the thickness of the thread accounts for the variation. More signatures might be advantageous, and a thin enough thread might make it possible. An increase in bulk at the spine, called swell, cannot be avoided altogether, however, and makes the rounded spine necessary (see the chapter on hardcovers).

Waxing the thread

Cut about a yard (1 m) of thread and pull it across a cake of beeswax to make it slippery and to avoid tangling. Thread a large sewing needle and you are ready to go.

Needle

Sewing

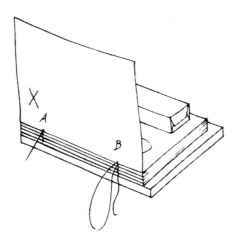

Place the stack of signatures on a board on the left of the worksurface, with the bottom signature facing up and the folds toward you. Starting with the topmost signature, open it to the middle, as shown in the illustration, and weight it. While holding the upright part, stitch through the fold at point A. Pull back the needle and sew through A again, this time from the outside, and bring it out again at B. Leave about 4 in (10 cm) of the thread hanging at A. Now take the next signature from the stack — the second to last in the book — and place it on top of the one just sewn.

Sew this and the following layers in the same way, pressing down the fold with the bone folder before the next one is attached. Never tie or knot the thread unless it breaks or runs out (see page 268 for instructions on the correct method of joining the ends). At first it may be difficult to achieve an even tension and to find the right points from the inside of the fold, since the markings are on the outside. It is important to keep the tension of the thread even, to align the layers carefully, to open the signatures precisely in the middle, and to sew exactly at the marked points.

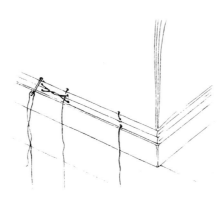

Another possible way of connecting the signatures is described in the chapter on hardcovers. For brochures the same method can be followed, omitting the tapes. Instead, as the drawing shows, the thread is crossed under the previous one. If there is too much swell, the process has to be repeated with thinner thread. Once you have some experience, you can predict the result after only a few signatures are sewn.

GLUING THE SPINE

Aligning

Adhesive

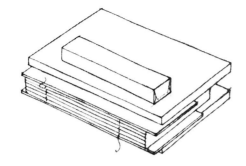

Align the sewn book-block and weight it between two strips of cardboard and two wooden boards. The cardboards should be flush with the stack, the boards set back slightly, to keep them free of adhesive. Apply pure PVA, or PVA with a small addition of paste, to the spine.

Aligning

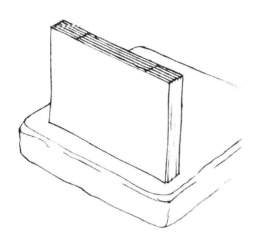

Remove the boards and realign the book-block at the front and the top.

Drying

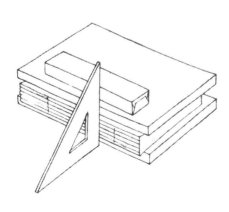

Weight the book-block again and let it dry between two boards for about 1 hour. The spine should stick out from between the boards and should be exactly square, since the position after drying will be permanent.

THE COVER

Material

The cover will be glued directly to the spine and has to be made from paper or board that can be folded easily. It should be medium-weight (120–160 gm^2) or heavyweight (up to 300 gm^2) for thick books. Good choices are wrapping or drawing paper, bristol board, or illustration board.

Grain direction

The grain direction must run parallel to the spine. Cut the cover, with a slight overlap for trimming. The width is two times the width of the book plus the width of the spine. Its height is the height of the book. Crease twice on the inside of the cover with a bone folder. The distance between the crease lines is the width of the spine.

Measurements

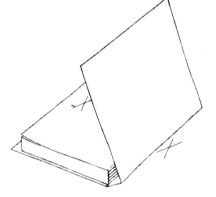

Titling

Try the cover for fit and trim the overlap. If you plan to put lettering on the spine, it should be applied now, while the cover can still be laid flat (see pages 284–287).

Gluing

Spread pure PVA on the spine and set the book-block onto the cover. If the adhesive layer is too thin, the bond will be weak; if it is too thick, the cover may warp. Check to be sure the book-block is correctly positioned, and rub the paper onto the spine. Then weight the book and leave it to dry.

Trimming

The front edges of the cover can be trimmed now. They should be slightly wider than the other edges or the same width.

Cover variations

Now the book is finished. Two possible variations are covers with one or three flaps. which can be folded around the first and last pages.

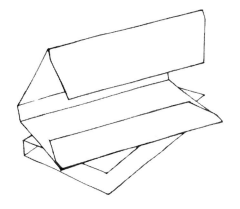

Another variation is to construct a cover without flaps, as before, and a second one with flaps, which will also be glued to the spine. The flaps are folded around the first cover and around the first and last pages of the book. This method makes it possible to use paper that otherwise would be too lightweight.

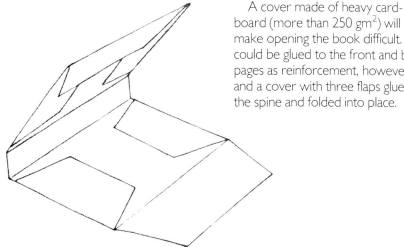

A cover made of heavy cardboard (more than 250 gm^2) will make opening the book difficult. It could be glued to the front and back pages as reinforcement, however, and a cover with three flaps glued to the spine and folded into place.

Checking the work

Errors and misjudgments can occur as you work, so stop frequently to critically assess your progress.

Are the stitches satisfactory?
Are the edges trimmed straight?
Is the book-block square? If not, the reason might be faulty alignment when the spine was glued, or that the block was disturbed when it was drying.
Are the overlaps of equal width?
Are there blemishes or traces of glue on the cover?
Is the spine concave or convex?

THE HARDCOVER BOOK

In a hardcover book, spine and covers are connected and attached to the book-block as a unit as the last step of assembling the book. This method contrasts with the one in which covers and spine are attached to the book-block in separate steps and covered afterwards. The hardcovers can be covered with paper, cloth, parchment, or leather. They can be flexible or stiff.

Some reflections seem appropriate at the beginning of this chapter. What is the cover? It serves the book, protects it, bears its title, and announces the content of the book itself. The thing that actually is being protected is the book-block, created of matter and spirit, big or small, long or wide, with a flat or rounded spine, made of lightweight or heavy paper, edges rough or smooth. The character of the book is influenced by the type, by the arrangement of letters and words on the page, by the symmetry or asymmetry of the titling, by decorations such as cartouches or illustrations. The very ideas expressed in the book need a fitting form.

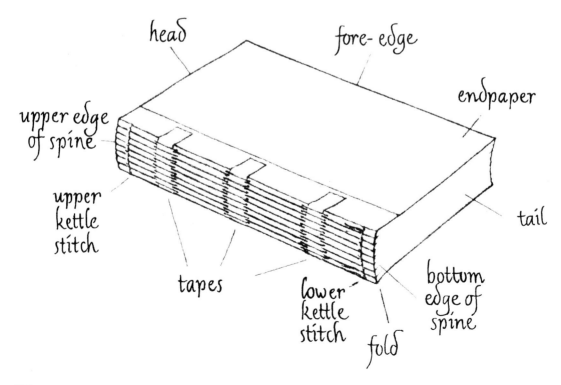

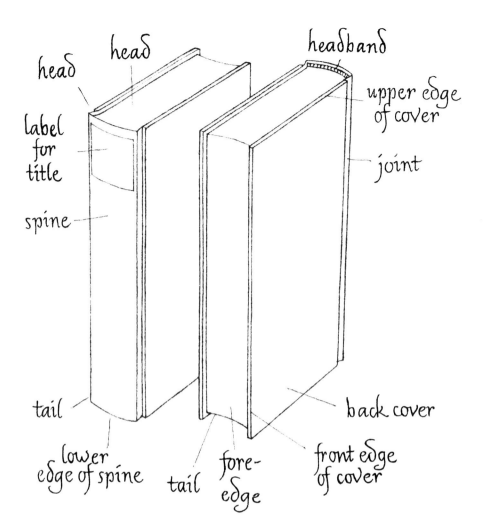

    The image of the book as expression of thought and form should guide our efforts. The cover design should be a natural representation of the book itself, with nothing forced or calculated about it. It is the job of the cover to prepare, not to overwhelm, the reader.

    The cover cannot be an end in itself, or it betrays its role, which is to serve the contents. Its dignity is diminished when it is designed to attract the reader with superficial charm. (Please note that I speak of the reader, not the owner.) This view is not supported by general opinion. It is nevertheless easily defended.

    Should covers be designed traditionally? Hardly anyone today could define the term. Let us be modern instead, modern in a candid way, without spelling out every detail: sparse rather than opulent. The shape of the book-block must never get lost in rounded forms and overhanging edges, in harsh contrasts, between inappropriate accents. Let us abandon unnecessary headbands and cords in favor of the functional and the suitable. This is tradition at its best.

    Work on a book should not be thought of as a two-part process, one mechanical and boring followed by another, creative and absorbing. All steps lead to the same end and form a unit.

PREPARATIONS

REPAIRING A BOOK

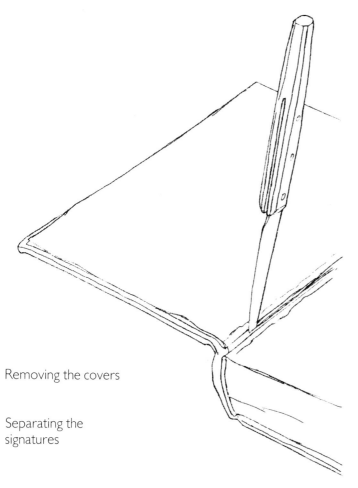

Removing the covers

Separating the signatures

You may be confronted with an old book in serious disrepair. Before its parts can be restored it has to be disassembled. Open the cover and the endpapers to an angle of 180 degrees and pull gently while holding the book-block in place. In the fold, next to the first signature, a space will become visible. If the book was machine-made, a gauze strip runs along that space. Cut it lengthwise with a sharp knife. In the unlikely case that you find tapes, it will be obvious that the book was hand-bound. Adhesive-bound and hand-bound books will not be discussed here.

After the tapes are cut, the book-block is freed from the cover and should be separated into its signatures. Under normal circumstances the first one will contain pages 1–16 on eight leaves, or, more exactly, four double leaves. In some books you will find small digits under the text on the first page of each signature, sometimes a corresponding one with an asterisk on the third.

These numbers are called signature marks. They aid the bookbinder in assembling the book.

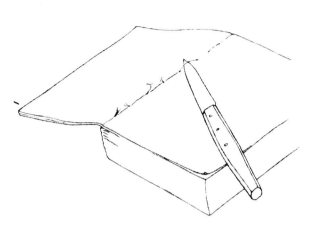

Before each signature can be freed, traces of glue have to be scraped off with the back of a knife or a bone folder. Between pages 8 and 9 the threads become visible and should be cut carefully without damaging the paper. Then close the signature and pull it off to the left. In some books, especially ones glued with PVA, the signatures are easier to remove to the right side. Proceed carefully — a pull in the wrong direction can tear the pages.

The following steps are repeated until all signatures are separated:

Turn the first signature to the left, face down; remove the glue from the fold of the next, open it, cut the thread, pull it off, and place it on the first one.

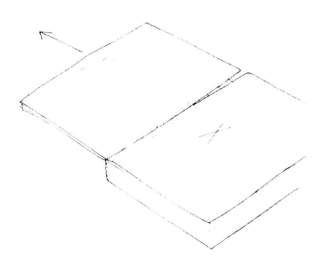

From time to time clean waste off the table. To protect the last page, leave the opened cover under the book-block until all layers are separated. In the end all the signatures should be in the correct sequence in a stack on your left.

Cleaning the signatures

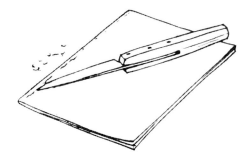

Recheck all folds for traces of glue and threads. Then align the stack and check the folds once more. The book-block should be in the same condition it was in before the original sewing. If you detect any tears, they have to be repaired now. Use tissue paper and thinned paste, applied with a watercolor brush to the edges of the tear.

Mending rips

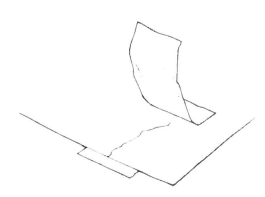

Align the edges of the tear, cover with tissue paper on both sides, and weight between strips of cardboard. After a drying period of ½ hour, pull off all tissue paper that is not glued. With some practice, this method makes almost invisible repairs possible.

If a leaf is torn in the fold, cut a ¼–½ in (5–8 mm) wide strip with the grain, the same length as the page, even if the rip separates only part of the fold. The reinforcement strip has to be lined up with the top of the page.

Mending folds

A weakened fold is treated in the same way. If no special mending paper or Japanese paper is available, a thin strip the color of the pages can be used. The aim is to make the patches as inconspicuous as possible.

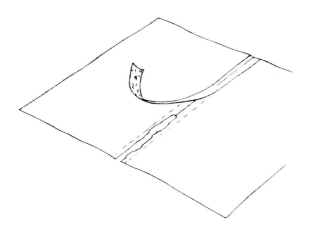

To fix a fold, open the double leaf, position it inside up, and align the two halves with a gap of about 1/16 in (1–2 mm) between them. If the leaf comes from the outside of the signature, the gap has to be appropriately wider. Apply adhesive to the reinforcement strip, center it on the fold, and set it on without stretching. Weight the pages and dry. After all repair work is done, check to see that all folds are in the right place and that the signatures are in the right sequence.

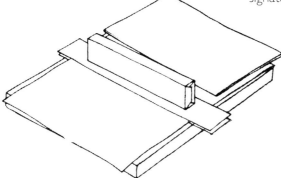

Restoring

None of the repair jobs described should be done for books of high value or bound artwork. Restoration of such materials is best left to the professional, since it requires special knowledge of the chemistry involved.

REINFORCING THE SIGNATURES

Sometimes there is a considerable tension to be felt when the cover of a book is opened. It acts between the cover and the endpaper, pulling on the threads of the first and second signatures, and often causes older books to split in this area. The remedy is to attach the first and second signatures to each other along a 1/8 in (4 mm) side strip and reinforce the outer layers of the two signatures along the folds.

Reinforcement strips

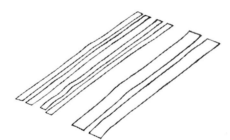

Choose paper in a matching color and cut strips that are the length of the spine and about 1/4–1/2 in (5–8 mm) wide. For small books paper will do; larger ones may require cloth. Cut two strips about 3/4 in (2 cm) wide for reinforcement of the outside of the first and last signatures after the endpapers are in place.

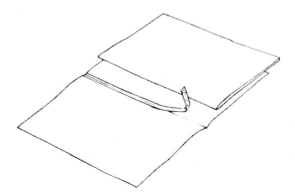

The outer pages of the first and last signatures are reinforced on the inner side. After the strips are glued in place, let the pages dry, refolded — but not too sharply — and put them back in place.

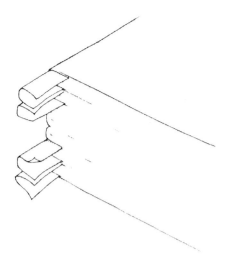

Do all repair work with paste, to avoid the danger of pages sticking together by accident. The ends of the reinforcements stick out at the top and bottom edges until they are trimmed. Afterwards it can easily be determined if a book was previously reinforced or not.

INSERTING PICTURES

To insert pictures or other loose leaves in a book, glue, or tip in, along an edge or, better yet, guard around a double leaf. When tipping in, apply adhesive to the back of the picture along its edge, set it onto the double leaf or signature, press, and weight (see drawing on page 259). To guard a picture page around a double leaf, apply adhesive along its edge and the edge of the double leaf in one step, as shown. Set half of a ½ in (8 mm) wide tissue guard onto the edge of the picture page.

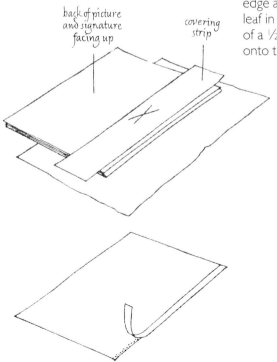

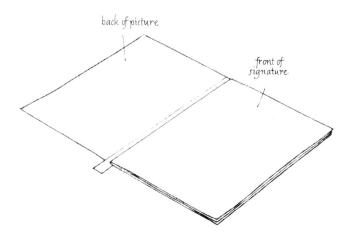

The other half is attached to the upper side of the double leaf. A slight gap, depending on the thickness of the signatures, remains between the edges. Press and weight.

ENDPAPERS

Endpapers are the visible connection between the book-block and the cover. Different types are possible. Most consist of a double leaf of white or colored paper, which is either attached along a narrow strip onto the first signature or folded around it with an additional strip, much as picture pages are attached.

The part of the double leaf that is glued to the inside of the cover is called the paste-down, the other one the flyleaf.

Choose a sturdy, medium-weight (no less than 100 gm^2) paper, depending on the format and the use of the book. Book (printing) and other tear-resistant papers can be used. The color can be either that of the book-block or one that eases the transition between book-block and cover. If the edges of the pages are to be colored later, this factor has to be considered too.

If the cover is monochromatic, marbled or otherwise patterned endpaper might be a perfect match. If you are in doubt, it is always better to be cautious rather than loud.

Choice of paper

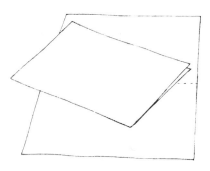

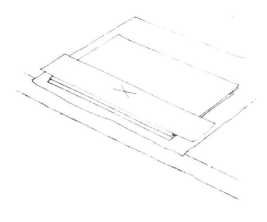

Arrange the two endpapers as shown in the drawing, exposing a ¼ in (3–4 mm) strip under newsprint, and apply PVA.

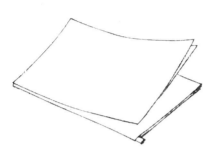

Set the endpapers onto the book-block. They should be flush with the book-block on all sides.

Outer reinforcement strip

A ¾ in (2 cm) cloth strip serves as reinforcement between the endpaper and the first signature. Apply adhesive to it and set the signature on so that about ¼ in (4–5 mm) of the width remains free.

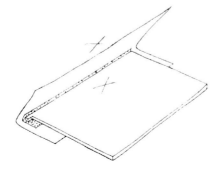

Fold the remaining part around the signature with the help of an extra sheet of paper.

Weighting

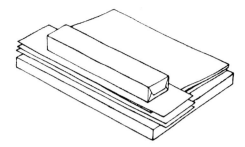

Weight the signatures between cardboard strips for 1 hour.

Assembling the book-block

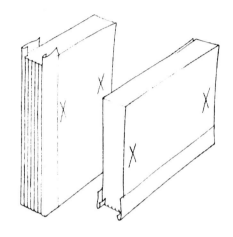

Add and align the first and last signatures with the rest of the book-block.

Weighting the book-block

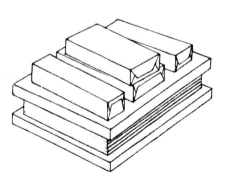

Leave the book-block weighted between boards for several days, then it is ready for sewing.

TRIMMING

Since no special equipment is assumed to be available, we trim signature by signature, as was done for centuries. The text on the printed page has to be taken into consideration before the pages can be trimmed. There are classical relationships between text and margins. They all have in common that the width increases proportionally, from the inside margin to the outside margin, and from the top margin to the bottom margin.

Considerations in text placement

Margins

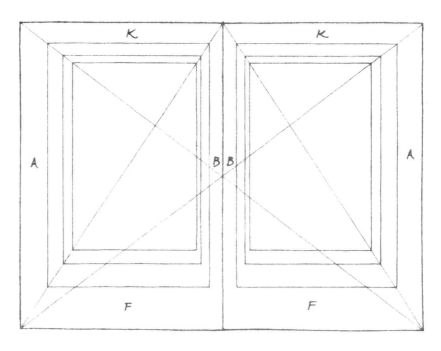

Proportions

Contemporary design

The drawing shows frequently used patterns for placing text on a double page. They can be adapted to pages of any size. Besides this classical design, there is the contemporary notion of balanced proportions, and there are numerous combinations of both views.

	There are also an increasing number of layouts that possess neither interest nor character. It is unavoidable that the intended proportions be changed through trimming. The original design should be preserved wherever possible, and the trimming should be limited to what is absolutely necessary.
Deckle edges Trimming Sanding the edges	Deckle edges too should be preserved where possible, even though generally more value is attached to them than is justified. Trimming the edges is a technique developed in the nineteenth century. The original deckle edges are cut back to the same size. Each signature is trimmed individually with the help of a straightedge. The cut protects the book-block from dust particles and makes it possible to color the edges. The cut edges can be treated with sandpaper for a fine finish. During sanding the book-block has to be pressed between two boards (see page 274 for coloring and page 242 for trimming.)

THE BOOK-BLOCK

SEWING ON TAPES

Spacing

Kettle stitches

For a spine of 6–10 in (15–25 cm) long, three tapes are enough; longer or shorter spines need more or fewer of them. The kettle stitches should be about ½ in (8–10 mm) from the ends of the spine if the book was trimmed before sewing. Kettle stitches are the chainlike structures (F in the drawing) that

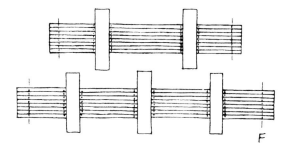

appear where the thread changes direction and leads from one signature into the next. The little knots result from the loops formed by two layers of thread.

Marking the positions

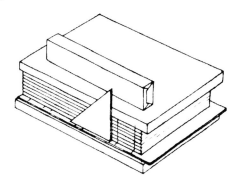

Mark the positions of the tapes on a cardboard strip and transfer them onto the spine. Only pencil should be used, never ballpoint pens or felt-tipped markers. Use the holes left by earlier stitches unless they require improper distribution of the tapes.

Tapes

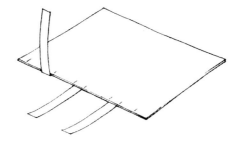

The ends of the tapes are glued under the edge of the cardboard. They can be made of various materials, but should be as flat as possible. Their length corresponds to the width of the spine.

SWELL

Thread

Swell and rounding

Swell and choice of paper

The thread has to be chosen with care. Its thickness determines how much the swell — the increase in thickness at the spine — will be, which in turn is responsible for the rounded back. The thread runs back and forth in the folds of the signatures. Its thickness adds up and expands the width of the spine.

Rounding the back shifts this stack of threads and compensates for the increase. Book-blocks with extreme swell tend to round out at the spine almost independently. The rounded spine, whether naturally developed or helped along with a hammer, makes the book stable and durable even if no further reinforcements are added to the spine.

The kind of paper used for the signatures determines how much the thread will change the width of the spine. Soft paper absorbs much of the swell; hard paper is not as accommodating. It follows that soft paper should be sewn with thicker thread than hard paper.

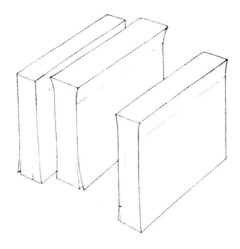

The left-hand book-block shows no swell at the spine. Soft paper and overly thin thread were used. The middle book-block shows too much swell, caused by the use of hard paper and thick thread or by sloppy or loose stitches. The right-hand book-block shows an ideal swell, which should be determined by the thickness of the cover boards and the desired shape of the spine.

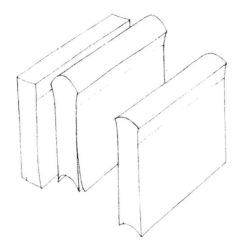

The left-hand book-block in the drawing here shows a convex-shaped spine and a corresponding bulge at the front edge, caused by lack of swell. The middle book-block, because of too much swell, is too rounded at the front. The book-block on the right, with a medium rounding, is the only one that preserves the image of the original book-block.

Swell and thickness of cover boards

The width of the spine should match the width of the book-block plus the thickness of the covers. Thick cover material requires a larger swell for a balanced appearance.

THE THREAD

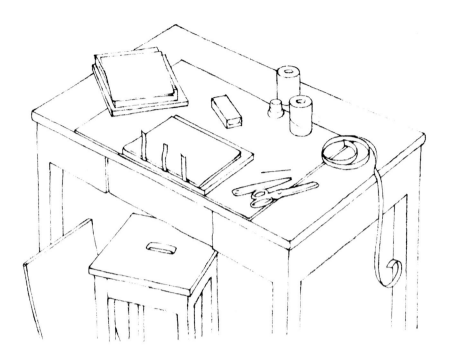

Several details have to be pointed out about the actual sewing process after you have checked the signatures for correct sequence. The thread is pulled over a cake of wax to diminish the danger of tangles.

The unstitched book-block is on the left, the first signature on the bottom of the stack. The last signature is already sewn. The first stitch is visible at the right kettle stitch and the loops around the three tapes.

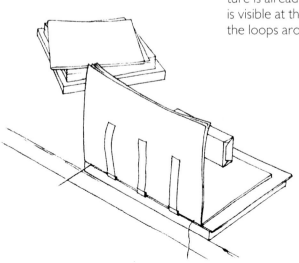

Direction of needle

Tightening the thread

At the first signature hold the needle horizontally; at the last one, point it down to avoid the endpaper, and at all other signatures hold it at a 45-degree angle. The path of the thread is shown in the drawings. It should be pointed out once more that each stitch toward the inside is made through a hole that was first poked from the inside out.

The thread has to be tightened at the very least when it reaches the kettle stitch. Between the first and the second signatures, and also between the second and the third, no knots are made at this point. Close the second signature, press it down with the bone folder, set the third signature on top, and open it. The thread enters the third signature at the kettle stitch, leads around the tapes, and comes out at the right kettle stitch.

Connecting the first and second signature

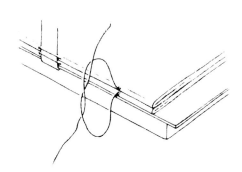

Tie the ends of the two threads . . .

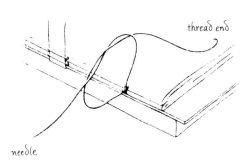

. . . then make the second half of the square knot and tighten.

Cut off all but ¼ in (5 mm) of the ends.

Holding down the folds

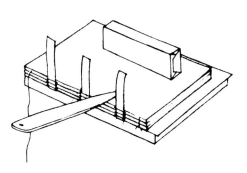

Press each signature down between the tapes with the bone folder after the sewing is done.

Upper kettle stitch

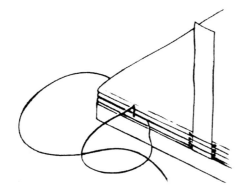

After the needle comes out at the left kettle stitch of the third signature, push the needle between the first and second signatures around the existing stitch and pull the loop tight.

Lower kettle stitch

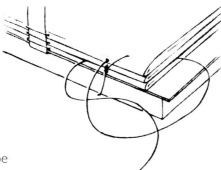

The drawing shows the same process at the right kettle stitch.

Keeping an even shape

If the loops are pulled too tightly, the spine will be narrower on top and bottom than in the middle. A similar result occurs if the signature folds are not pressed down sufficiently with the bone folder after sewing.

Knots

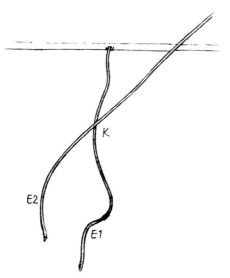

At the end of the thread, or after a rip occurs, the new thread is attached next to a tape on the outside. No loose ends should ever be visible inside the book. The end of the new thread, $E2$, is placed over the end of the old one, $E1$, forming the crossing point, K, which you hold between thumb and index finger of your left hand.

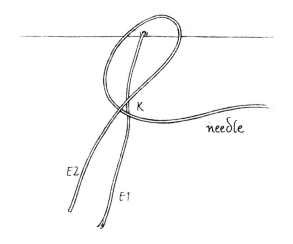

Form an underhand loop with the new thread and hold that loop too with your left hand.

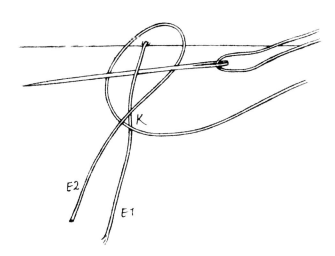

The new thread is then sewn between the book-block and point K, underneath the old thread.

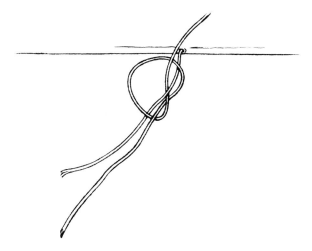

While you continue to hold the ends loosely with your left hand, with your right hand pull up the new thread until it rests against the paper. Pull the knot tight.

After the last signature is sewn (it is the first signature of the book), loop the thread twice around the kettle stitch and cut off the end, leaving ¼ in (5 mm). Free the tapes from the cardboard, check for even stitches, regular patterns around the tapes, and even tension at the kettle stitches.

Checking tension

Gluing the first
and second signatures

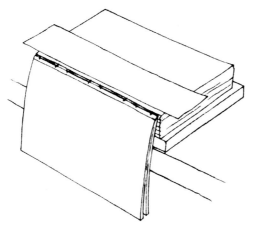

Shake the book-block to make sure all the pages stay in place, and check for flush edges. Many errors still can be corrected before the spine is glued, and a repetition of some steps is far preferable to an imperfect end product. Now all that is left to do is to attach the first and second and the last two signatures to each other with adhesive applied to a ⅛ in (3 mm) strip of the edge.

The drawing shows how the first signature is covered by a protective strip, leaving a narrow margin free next to the fold. Apply PVA, close the first signature, and press. Repeat these steps with the last two signatures and weight the closed book-block for 5 minutes.

Regulating the swell

The swell can be reduced slightly with gentle strokes of a hammer. The front of the book-block has to be pressed together during this process to make sure that the spine does not round yet. Before the spine is compressed with the hammer, attach the first and last signatures to each other; glue the ends of the tapes afterwards. The hammer should hit the book-block at a right angle.

Gluing the tapes

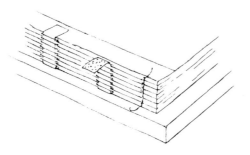

The ends of the tapes should end about ⅛ in (2 mm) before the edge of the reinforcement strip. Glue the ends onto the strip with PVA.

GLUING THE SPINE

Aligning

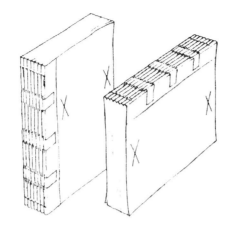

Align the book-block at the top and the side, but not at the spine, to even out the front. If the book-block were to be machine-trimmed, the spine would be aligned before and after gluing.

Gluing

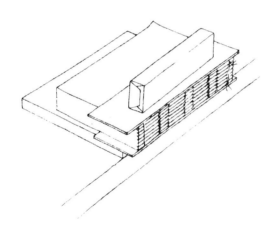

Place cardboard strips under and over the book-block, flush with its edge, and weight them. Then apply a generous layer of PVA in strokes from the outside in, to avoid any traces of adhesive on the edges.

Aligning

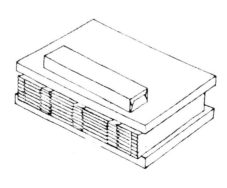

Drying

Align the book-block again at the top and the side and take care to put it down on the work surface between two boards without any twist disturbing the alignment, because it will keep its form permanently after drying. The spine should stick out ¼–½ in (5–10 mm) from the board. Then round the spine.

ROUNDING THE SPINE

Rounding the spine is not done just for aesthetic reasons. The appearance of a book often would be more pleasing with a flat back. We observed that a sufficient swell causes the spine to bulge out. A rounded spine increases the book's stability.

Only book-blocks less than ¾ in (2 cm) thick should be left unrounded. If you do not plan to round the spine, keep a little space between the covers and the spine. Books with rounded spines appear more traditional, while flat spines seem more modern, an effect that has to be considered during the planning stages, when you determine the character of the book.

A slight rounding of between a quarter to a third of a circle's circumference is often an acceptable compromise. During further handling, the rounding often diminishes, so make it slightly more pronounced than seems necessary.

Rest the book-block on a solid surface and hold it in place with one hand, thumb on the front edge, fingers on the block. Hold a medium-size hammer with the other hand. In the illustration the bookbinder would be standing behind the book-block. The side of the hammer should hit the spine in a series of even strokes, moving back and forth. The outermost signatures should not be hit directly, because they could move out too far.

Ideal shape

Rounding

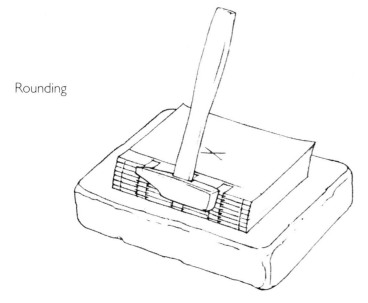

Keep the spine in contact with the work surface, and turn the bookblock upside down occasionally to ensure even progress. It is essential to check your progress frequently.

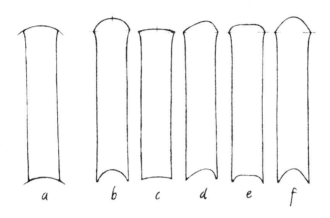

Right and wrong rounding

Drawing *a* shows a perfect rounding. Drawing *b* is even but too pronounced; *c* even but too flat. The spine would flatten after the book was opened for the first time. Drawing *d* shows an irregular rounding on one side, *e* on the outside, and *f* in the middle.

Pressing

With the help of a hand press, the shape of the spine can be stabilized further.

COLORED EDGES

White edges can be beautiful, and color should be introduced only if it was included in the overall design. Page 209 explains how to prepare edges of books with individually trimmed signatures for the coloring procedure. The book-block is weighted between two boards, so that no color can seep between the pages. The spine faces left.

No color is far better than a poorly chosen one, and the color has to be mixed very carefully to match the rest of the book. Trial applications on strips of paper are an absolute necessity.

Mix the paint with just enough water to make a smooth application possible. If it is too thick, a crust will form; too thin and the color will leak into the interior of the book-block. A single application of paint has to suffice.

To keep the spine clean, the brushstrokes should go from the spine to the sides. Leave the weights on the book-block for about 5 minutes; then the paint should be dry. If waves appear along the edge, there was too much water in the paint mixture, in which case leave the weights in position for a longer period.

When the paint is completely dry you can smooth it with a soft brush or a thin layer of beeswax.

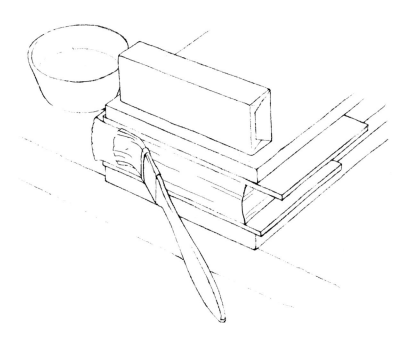

THE HEADBAND

Headbands are the little strips of fabric often visible at the top and bottom ends of the book-block at the spine. Originally a functional part of a book's construction, they became a decorative element during the course of centuries. They still keep dust particles from entering the space between the back of the book-block and the spine cover.

Except for providing an accent of color where it fits the design of the book, headbands are dispensable, since there should be no need to hide the top and bottom edges of the book-block.

Exquisitely hand-tooled headbands can be found in volumes of parchment and leather, and machine-made ones come by the yard. Neither is appropriate for our purposes. A handmade headband can be matched to the book and is therefore the best choice.

Cut a piece of cotton cloth, plain or striped, twice as long as the width of the spine, and a narrow strip of cardboard, parchment, or leather. For delicate headbands use a length of thick thread. Apply PVA to the cloth or the cardboard strip and set the strip onto the cloth. Fold the cloth over as shown in the illustration. After it is pressed down and dried, cut the headband into two halves exactly the width of the spine.

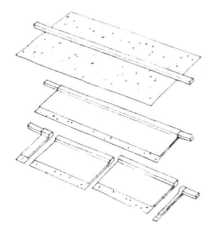

The drawing shows how the headbands should be positioned in relation to the edge of the spine. They reach the kettle stitches on each side, without touching them. The edges of the spine cover are slightly higher than the headbands, but the cover determines the position of the headbands, not vice versa.

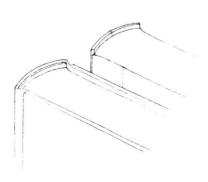

REINFORCING THE SPINE

Reinforcing the spine between the tapes makes the spine sturdier and helps it hold its shape. It is done with strips of paper or, for books that need greater flexibility, with cloth. Medium-weight (about 100 gm^2) writing or wrapping paper is a good choice.

If you anticipate heavy use, use double or even triple layers of the material, but never reinforce the spine until it is so rigid that it will keep its form forever or until it breaks apart in the hands of a terrified reader. Books are meant to be read, and their construction should serve this purpose.

Cut the reinforcement strip with the grain direction to the exact width of the spine. Cover it with paste to soften the material, and then apply a layer of PVA to the strip as well as to the spine.

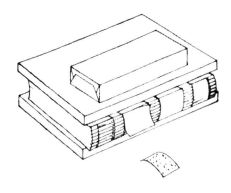

Set the strip on the spine starting at the right kettle stitch. Rub it on with your fingers and the bone folder. This should be easy to do since the paper was presoftened with paste. At the edge of the first tape trim the paper strip so as to leave the tapes free, and continue gluing reinforcements until you reach the last tape on the left.

The second layer of reinforcement covers the entire length of the spine, including the headbands. Unless you rub these strips on very carefully, you defeat their purpose.

Check the shape of the bookblock once more before allowing it to dry.

THE CASE

CONSTRUCTION

Whether the covering of the case is to be paper or cloth, its basic construction does not change. It consists of two cover boards (*a*), a spine (*c*), and a connecting paper strip (*b*). All pieces are cut with the grain and identical in height.

The drawing serves as a reminder and illustration of my earlier statement that the outer surfaces of the covers should be in the same plane as the outer folds of the book-block at the spine. The swell, of course, was determined by the choice of thread.

Cut the cover boards. In the finished book they should overhang all sides equally, but cut the front cover ½ in (10 mm) wider at the front edge at first.

The width of the overhang is integral to the appearance of the book. It should be measured with moderation, to reflect the shape of the book-block. A 3:2 relationship between the width of the overhang and the thickness of the boards is better than a 1:1 ratio.

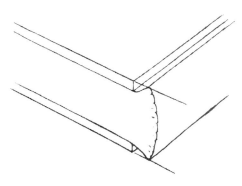

Cover size

Spine

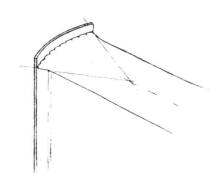

Cut the spine cover from heavyweight (200–300 gm^2) bristol board or mat board in the width of the spine. Take the measurement with the help of a paper strip that follows the round shape of the spine. A flat spine requires a thicker spine cover.

Connecting strip

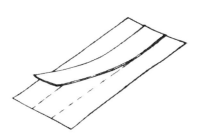

Cut the connecting strip from medium-weight (about 120 gm^2) paper. It should be 2 ½ in (6 cm) wider than the spine cover.

Combine the two parts after they are matched in height. Use PVA with a small addition of paste.

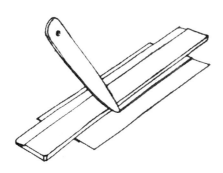

Turn the piece upside down and rub by hand and with a bone folder. Mold the paper to the side edges of the cardboard strip.

Spacing

For the case to function properly, the distance between the cover boards and the spine cover must be right. Try ¼ in (5 mm) as a usable average. Reduce to ⅛ in (3 mm) for lightweight boards and increase it to ⅜ in (8 mm) for thicker ones. Another factor to be considered is the thickness of the covering paper or cloth. Experience will be your only reliable guide.

Mark the distances on the connecting strip, especially during first projects.

Applying adhesive

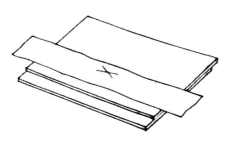

Apply PVA to a ½ in (1 cm) wide area of the two cover boards, using the technique of gluing series.

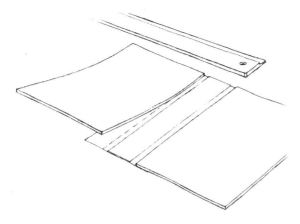

Set one cover board after the other onto the connecting strip and check the alignment with a straightedge. It is important to note that the spine cover is glued to the inside of the strip, but the cover boards are glued to its outside.

With a knife and straightedge trim the edges of the connecting strip that are glued down. Never try to tear off the excess.

Right and wrong spacing

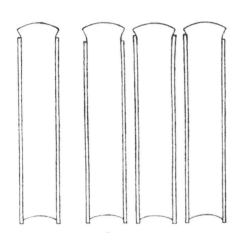

The drawings show correct and incorrect placement of the cover boards. From left to right: normal, too far from the spine cover, too close, and unequal.

Trimming

Rounding the spine

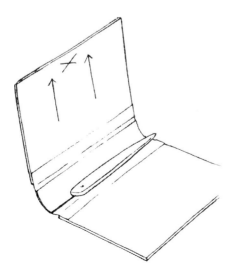

Now trim the front edges. They should be about 1/16 in (1 — 2 mm) wider than the top and bottom edges to accommodate possible shrinkage of the connecting strip at the spine and in the joints.

To fit the case properly to the book-block, you must round the spine cover. Place one cover board on the work surface. Hold up the other board and move the bone folder from one edge of the spine cover to the other, until you achieve the desired shape.

Inserting the book-block

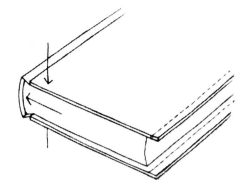

Now put the book-block into the case. If all parts fit, including the back, mark the front edges of the book-block on the insides of the cover boards; mark the inside of the front cover with an F to match the F that marked the front of the book-block. This step will help you avoid combining the two parts incorrectly later on.

Edges

After trimming the cover boards to their final size, check the fit of the book-block, the width of the edges, and their relationship to the book-block one more time.

Treat the edges of the cover boards with sandpaper if the cutting has left them rough.

COVERING THE CASE

Half binding

The first method described is the half binding, which presents fewer difficulties to the inexperienced craftsman than binding an entire book in cloth.

Proportions

Start by determining the relationship of paper area to cloth area, which greatly influences the character of the book. Materials, colors, and textures must go together. In extreme cases nearly the entire board can be covered with paper, or the relationship of cloth to paper on the board could be 1:3. Mathematical proportions are often helpful guides. The chapter on portfolios contains advice about colors.

Portfolio and book spines

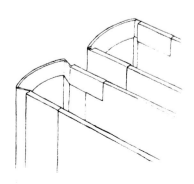

Covering the spine of a book is very similar to covering the spine of a portfolio. The main difference is that with a book you must push cloth into the joints from the outside and work out a profile with the bone folder. The drawing shows a book in front, a portfolio in back.

280

All-cloth cover

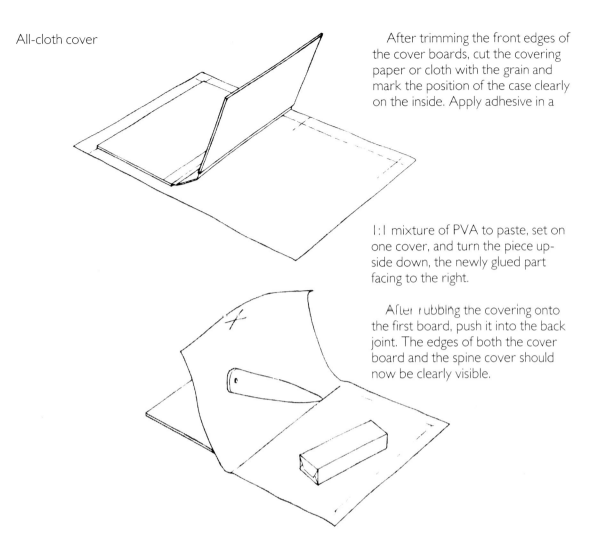

Fitting the case

After trimming the front edges of the cover boards, cut the covering paper or cloth with the grain and mark the position of the case clearly on the inside. Apply adhesive in a 1:1 mixture of PVA to paste, set on one cover, and turn the piece upside down, the newly glued part facing to the right.

After rubbing the covering onto the first board, push it into the back joint. The edges of both the cover board and the spine cover should now be clearly visible.

Without further delay glue on the other cover board, rub, and turn over again. Miter the corners at a 45-degree angle and fold the overlaps (as described earlier in the chapter on folding). After rubbing down and checking all areas, edges, and corners, trim the overlaps so they are equal in width.

Following the previous method, again fit the case to the book-block.

CONNECTING THE CASE AND THE BOOK-BLOCK

If a press is available, use it to make this connection. Apply adhesive to the endpapers and attach them to the covers with enough pressure to prevent the formation of wrinkles. If you do not have access to a press, use weights. The connection of case and book-block is best done before the case dries completely.

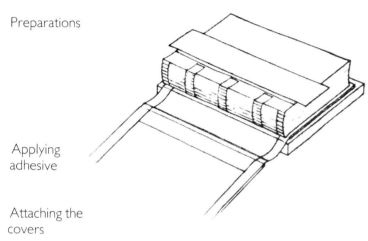

Preparations

Applying adhesive

Attaching the covers

Fit the book-block into the case, place everything on a wooden board, and open the front cover without moving the book-block. Cover the top page with newsprint, leaving a 1 in (2 cm) wide strip at the back edge, and apply a thin layer of PVA, without spreading any of it onto the spine or the sides of the book-block. Remove the paper strip and close the cover. Immediately afterwards turn the book over and repeat the process for the other cover.

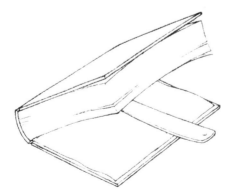

It is essential to rub the glued strip from the inside with your fingers and bone folder, while holding the covers open to an angle of no more than 40 degrees.

Weighting

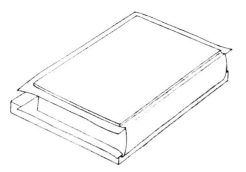

Insert thin blotters under each cover and press or weight the book for 15 minutes. Then cut the attached leaf at the adhesive line, remove the rest, and press the cut edge smooth with the bone folder. This leaf is no longer needed.

THE PASTE-DOWN

Lining

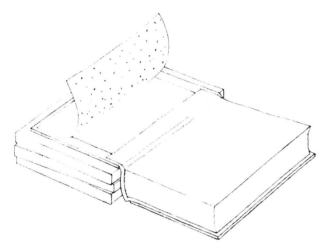

Line the inside of each cover board with a paste-down, cut with the grain. The paste-downs are the same length as the cut endpaper and reach the inner edge of the cover board, but do not extend into the joint. For this last step use an adhesive mixture of PVA and paste in a 4:1 ratio to prevent too much tension inside the covers.

Drying

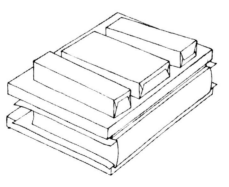

Do not consider the work completed until everything is dry. Weight the book overnight. Exchange the blotting papers for fresh ones after 1 hour to keep moisture from getting into the book-block, and place clean sheets of paper between the book and the wooden boards.

TITLING

No book seems complete until it bears a title on its spine. Titling is the book's face and makes it recognizable to readers other than the owner, who might identify it by shape and color. Titling also ensures that the book will be placed on the shelf right-side-up.

The following remarks remain valid whether the title is done with typography, by hand, or with the help of a press. Professional bookbinders usually choose the latter.

Where to place the title? If you choose a traditional type style, its placement should follow traditional patterns.

To make reading easier, bookbinders often set the title on the spine horizontally, for easiest reading when the book is on the shelf. Narrow spines preclude this option; titles arranged sideways always run from top to bottom. It may be necessary to shorten the full title if space is limited, but the result should always seem complete and contain no apparent reductions or abbreviations of words.

You can use the following measurements as a guide: Subtract ¼ in (5 mm) at the top and ¾ in (15 mm) at the bottom from the length of the spine, and divide the rest into six equal parts. The rectangle second from the top should bear the titling. Alternatively, divide the spine into four parts.

Modern typefaces, whatever they may be, do not carry the same restrictions as traditional ones. A personal sense of proportions is the only available guideline.

When is a label on the spine advisable? When the cover material precludes direct titling, or if the added color of the label enriches the design. You can choose from a variety of colors, shapes, and arrangements for labels. Consider the color contrast of label to cover material. White is rarely a good choice, especially if the cover is dark. Gray or yellowish tones are often most satisfying, but a different nuance can influence the character of the book and give it a needed accent. Again, plan the total image. Harsh contrasts rarely create a harmonious effect.

Make the label a little narrower than the spine to protect the edges of the label from damage.

With tempera paint you can produce any hue desired for the label. After painting, wax and polish with a soft cloth, or leave the label unpolished if this seems fitting. Attach the label with PVA, either by pulling it across a glued surface or by applying the adhesive directly with a finger.

The design and titling of the spine should match the type used in the text and the general ambience of the book. Romantic novels and nonfiction must be treated in distinctly different ways.

The words can be written with a ballpoint pen, fountain pen, felt-tipped marker, or even a pencil. (Pencil lines have to be protected from wear; apply fixative before the label is attached to the cover, since it may discolor the paper.)

Good results do not depend on the medium so much as on a feeling for form and an understanding of limits. Our grandmothers labeled jelly glasses and eggnog bottles with natural simplicity. In our work we should emulate this spirit, avoid any forced cuteness, and strive for aesthetic principles. Neat handwriting is better than stenciled lettering.

Titles written in calligraphic scripts such as italic, especially if executed without personal touch, have to be matched carefully to the type inside the book.

You can write titles in upper- or lowercase letters, upright or with a slant, or in combinations of the above. The author's name can be slanted, the title straight, or vice versa. Letter sizes can vary; for instance, the author's name can be written smaller than the title.

Labels typed by machine or drawn with templates may not be right for every volume but can add interest to certain cover designs. Larger volumes often benefit from big and mechanically produced lettering.

Rub-on letters should be used only with caution. Their appearance can rarely be reconciled with the handcrafted covers that they should enhance. Cloth-bound volumes should never be titled in this way, because the surface of the fabric is usually too rough to make a durable application of such letters possible. Cost can be another prohibitive factor.

The examples show different combinations of titling and should stimulate interest in further variations.

Spontaneity, if it is not confused with sloppiness, is an asset that should not be hidden by too many corrections. A book's title should be a design in its own right, whether it appears lightly written or meticulously constructed.

PHOTOGRAPH ALBUMS

Store-bought photograph albums are usually traditional in their case style. Simulated leather bindings, padded covers, and large spines are their hallmarks, and they seem most unlikely companions for such modern products as photographs.

Designs that emphasize clean edges are called for, with lightly rounded spines and narrow overlaps on the covers. Large photographs can be stored in portfolios or hinged boxes (see page 158).

I will describe a side-sewn album with or without spine, an adhesive-bound album with a flexible spine, and one sewn through the folds, all with covers.

Photographs are not the only collectibles that may be housed in such a book. Others examples are postcards, clippings, and all kinds of swatches.

Do not choose a format indiscriminately. Arrange the photographs on the sheet in groups until you find pleasing proportions. During this process you can cover the edges of the paper with strips of dark paper to simulate different formats. If you have no size preference, use simple fractions of existing sheet sizes as economical guides.

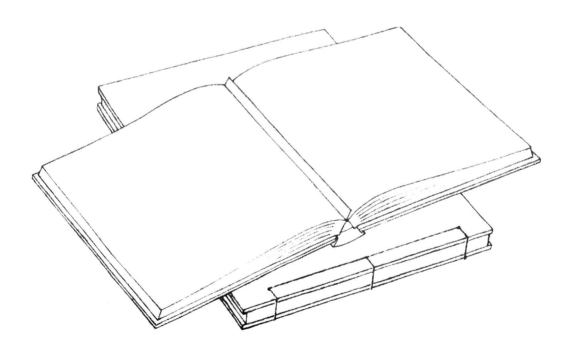

A SIDE-SEWN PHOTOGRAPH ALBUM

Material — Mat board is available in white, off-white, black, gray, and many other colors. Use the two-ply weight (between 220 and 400 gm^2). White backgrounds or grays with a blue undertone often complement photographs best.

Format — The type of binding used for the following album makes part of each page unusable. Therefore, add 1 in (2 cm) in width to the desired page format. Sewing signatures will diminish the usable area as well.

Trimming — It is difficult to trim photograph albums after they are bound. Gluing and sewing require careful attention, therefore, since slight errors cannot be corrected if the pages are already cut to the final size before sewing.

Stitching — Once you decide upon the format, consider the distribution of the stitches (see pages 220 and 224). Then cut an equal number of pages and folds.

Cutting — Cut the pages to the desired height and width. Next cut shims, which will compensate for the thickness of the photos. Check for correct grain direction and right angles.

Creasing guide

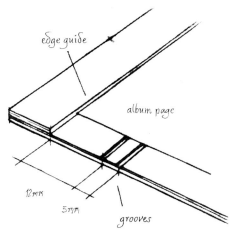

The illustration shows the special construction of a guide for photograph albums.

Shims

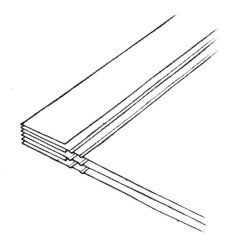

Crease along two parallel lines ¼ in (5 mm) apart. The first crease is about ½ in (12 mm) from the edge.

Gluing the block

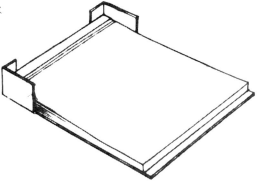

Construct a second guide, for the alignment of the book-block, by attaching the pages to each other with dots of PVA.

Punching holes

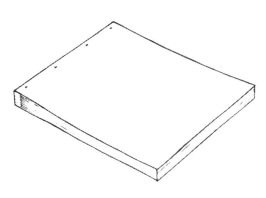

With the aid of a cardboard template, punch holes in the pages over a solid work surface. Hold the awl vertically, and make the holes no larger than necessary. You may have to punch them in consecutive operations.

Cutting covers

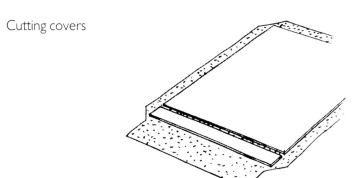

Cut gray cardboard with the grain for the two cover boards and two ½ in (10 mm) wide strips, which will form the joints together with the covers. The distance between the two parts of each cover is two times the thickness of the cardboard. The complete covers will be flush with the book-block at the back and overhanging slightly on the other sides. Use bookbinder's cloth for covering material. You can cover the boards completely or partially; the rest should then be covered with paper. Glue with a 1:1 mixture of PVA and paste (see text on adhesives and folding). The illustration shows how to work the cloth into the joints.

Covering the covers

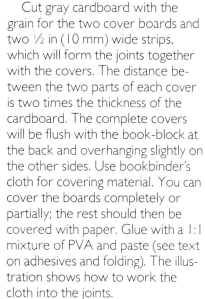

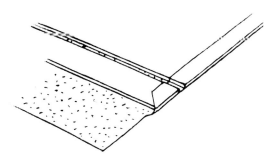

Rub the cloth into the joint only on the inside.

Lining

Now trim the overlaps to equal width and line the covers with medium-weight (120 gm²) paper, using the same mixture of adhesives as before (see also page 294). If you need a greater tension, add more paste to the mixture.

The lining should end about ⅛ in (2–3 mm) before the joint and have the same dimensions as the book-block. After the covers dry, punch

Drying

Punching holes

holes from the outside and secure the covers to the book-block with several drops of PVA.

Sewing

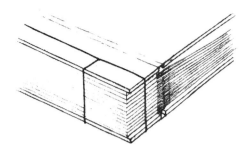

The stitches are more than decoration. Without them, the bookblock and the covers would come apart. For variations of stitches, see pages 219–224.

Photo album with sewn book-block and case

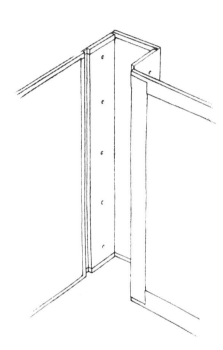

To construct connected cover boards, cut a spine cover from the same material as the front and back covers. Leave a distance of two times the thickness of the cardboard between the parts. After covering the outside with cloth, reinforce the inside of the spine cover with cloth cut wide enough to overlap onto the inside of the front and back covers. After applying adhesive, rub the cloth into the joints. You can cover only the spine with cloth, while covering and lining the front and back cover boards with paper, or you may cover all with cloth.

AN ADHESIVE-BOUND PHOTO-GRAPH ALBUM WITH A COVER

Creasing

Creasing guide

Gluing

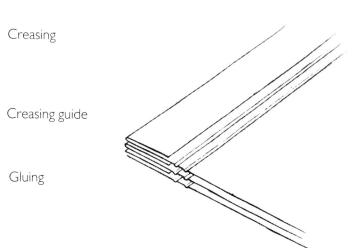

In this album each page has a ½ in (10 mm) wide fold. Crease along two lines. This will make it easier to open the album. Adjust the format accordingly.

You also need two double pages for the ends of the book-block and endpapers, which can be made simply by putting the sheets on a soft surface with the help of a guide as described on page 290. The parallel creases should be 1/16 in (1–2 mm) apart. The first crease is about ¾ in (22 mm) from the edge of the paper.

Attach all sheets or, better, all folds, to each other. The double leaves for back and front need attached folds to serve as connection parts to the book-block.

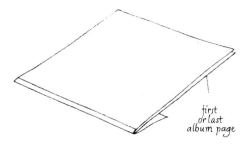

first or last album page

Cut the strips from the same material as the cover cloth, in the length of the spine and about 2 in (5 cm) wide. Glue them onto the folds of the double leaves along a width a ⅛ in (4 mm). Apply adhesive on the right side of the material. After a 10-minute drying period, fold the strips sharply around the edge and rub them down.

Attaching double leaves

Covering the spine

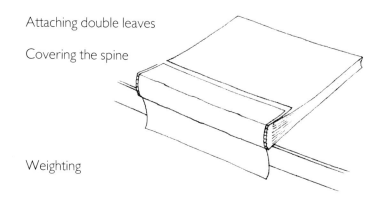

Folds facing out, glue the double leaves onto the book-block along ¼ in (5 mm) wide strips. Cover the back with Japanese paper or cloth. The strip used should be slightly shorter but 1 ½ in (4 cm) wider than the back. The overlap will reinforce the folds attached to the endpapers. Leave the book-block unweighted, inserting thin blotters under the folds to keep moisture out of it.

Weighting

Covers

The construction of the covers and variations are explained on page 277. For a flat spine, choose a material two-thirds the cover thickness. The text also explains how to connect the cover and the book-block. Only one significant difference exists: After fitting the book-block into the case, in one instance you remove the outermost layers almost entirely; in the other instance, only along a very thin strip. The lining of the covers might be considered a deviation from the earlier pattern, because photo mat board may not provide enough tension when used as a lining.

Lining

Attach writing or wrapping paper (about 120 gm²) underneath with the same adhesive as before. If cut with precision, this layer can also even out the height difference caused by the cloth overlap. Weight and dry the covers flat.

After 6 hours, when the effect of the first layer of lining becomes apparent, install the second lining of photo mat board — with PVA if the covers are perfectly flat, otherwise adding a small amount of paste to the adhesive mixture.

Drying

A special drying procedure, necessary because the covers do not lie flat on the book-block with its folds, is described on pages 299–300.

A SEWN PHOTOGRAPH ALBUM WITH A COVER

If the spine is wide in comparison to the rest of the book-block, and especially if the spine is very round, the album seems plump. Limit to the absolute minimum, therefore, shims needed to accommodate the added thickness caused by inserted material.

Do not use flat spines for books over 1 in (2 cm) thick. If you choose a flat spine, the distance between back and covers has to be wide enough to make opening the book easy.

Sewn photograph albums should consist of signatures of two or three double leaves and separate shims, cut from the same material. In addition, you need a double leaf each for front and back and two single leaves for liners.

Cut the sheets in the desired format, crease them once along the middle and once parallel to the middle on each side at a distance of about ½ in (12 mm) from the middle. Cut all pieces in grain direction.

To produce a guide for easier creasing is time well spent (see page 36). The drawing below shows the pattern.

Shims

Flat spine

Signatures

Cutting the sheets

Creasing guide

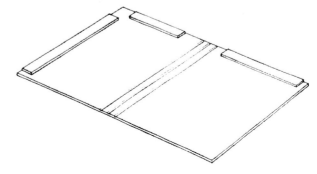

Shims

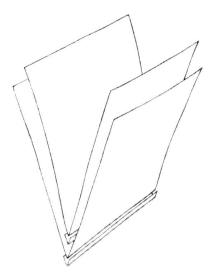

The drawing shows a signature consisting of two double leaves and two shims, both ½ in (1–2 cm) folded width, and the same length as the pages Skip the strip on the outer layer of every other signature to avoid double layers of strips between signatures.

Fold each shim like this: Fold a sheet, preferably cut from the same paper as the page itself, in the middle, and trim it afterwards. You need two strips made from cloth-covered paper to connect the book-block and the cover. Anchor the strips under the double leaves sewn to the front and back of the book-block.

Connection strips

Cut 2 in (5 cm) wide connection strips that are slightly longer than the spine. Use the same cloth as for the cover and medium-weight paper.

Attach each paper strip to a cloth strip with a 1:1 mixture of PVA and paste. Weight them and let them dry for 2 hours.

For clarity I have described this step here, but the work should be done at the start so that the connection strips can dry while you are preparing the pages. After the strips are dry, trim them to size and glue them to them to the edges of the double leaves along a strip ¹⁄₁₆ in (1–2 mm) wide.

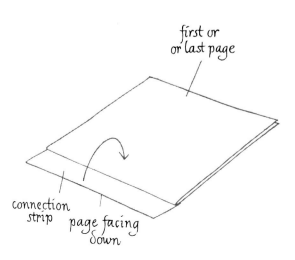

Apply PVA onto the fabric side, covering the rest of the strip. When the connections dry, fold the strips around the edge of the double leaf and add the pieces to the book-block. Align and press the still unsewn book-block between wooden boards for several days.

Sewing

Tapes

Sew the signatures in the same way as in a book. The number of tapes depends on the length of the spine (see appropriate chapters for details.) Sew the book-block, and glue down the ends of the tapes.

See the text on gluing and rounded spines on pages 271–273.

Reinforcing the spine

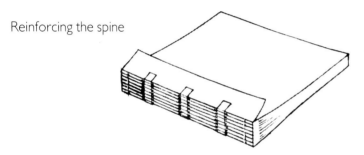

Reinforce the spine with thin cloth strips between tapes; then allow it to dry overnight.

See pages 277–281 on how to construct the case.

Case

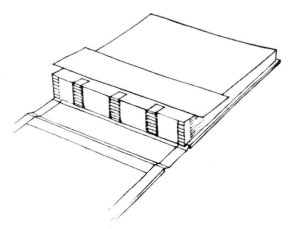

Place the book-block onto the opened case. Cover the front connection strip with newsprint except for a ¼ in (5 mm) wide strip. Apply PVA to the exposed area and remove the newsprint.

Rubbing down the
connection strips

While holding the book-block in position, close the cover and press it down. Small corrections for fit can be made now. Immediately afterward repeat the procedure for the other side. Rub on the connection strips from the inside (see page 282). Do not open the cover farther than necessary. Weight the book and let it rest for 10 minutes.

Trimming the
connection strips

Now trim off the connection strips at the adhesive line. During this and the following steps, support the opened covers with boards of appropriate thickness.

Lining

Line the covers as described on page 294. The lining paper should be cut the same height and width as the book-block, just a little smaller than the covers to allow an even border all around.

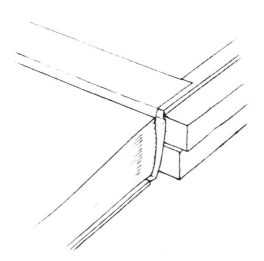

Drying

Weighting

In a book with sewn-in shims, the covers do not rest on the pages with equal pressure, so they must be dried and weighted in open position. The book-block remains vertical during this period.

Another solution is to suspend the hanging book-block between two tables and weight each half of the cover on one of them. A third possibility is to equalize the height differences of the book-block by inserting extra sheets and pressing it in the usual manner. Insert blotters under each cover, set wooden boards on the outside, leaving the back free. Weight the book and allow it to dry overnight.

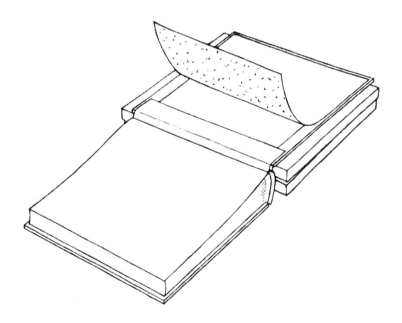

BIBLIOGRAPHY

Bosch, Gulnar; Carswell, John; and Petherbridge, Guy. *Islamic Bindings and Bookmaking*. Chicago: The Oriental Institute Museum, 1981.

Cockerell, Douglas. *Bookbinding and the Care of Books*. New York: Taplinger, 1978.

Da Vinci, Leonardo. *The Madrid Codices of Leonardo Da Vinci*. Translated by Ladislao Reti. New York: McGraw-Hill, 1974.

Davenport, Cyril. *The Book: Its History and Development*. Detroit: Gale Research, 1971.

Diehl, Edith. *Bookbinding: Its Background and Technique*. New York: Dover, 1980.

Dürer, Albrecht. *Unterweisung der Messung*, 1525 Facsimile. Jošef Stocker-Schmid, 1966.

Goldschmidt, E. P. H. *Gothic and Renaissance Bookbindings, Exemplified and Illustrated from the Author's Collection*, 2 vols. 1928. Reprint. Amsterdam: N. Israel, 1969.

Haüy, Rene Just. *Traite de Mineralogie*. Paris: 1801.

Hunter, Dard. *Papermaking: The History and Technique of an Ancient Craft*. New York: Dover Publications, 1978.

Ikegami, Kojiro. *Japanese Bookbinding: Instructions from a Master Craftsman*. Adapted by Barbara Stephan. New York: Weatherhill, 1986

Johnson, Arthur W. *The Thames and Hudson Manual of Bookbinding*. London: Thames and Hudson Ltd., 1981.

Jugaku, Bunsho. *Paper-making by Hand in Japan*. Tokyo: Meji-Shobo, 1959.

Kapr, Albert. *The Art of Lettering: The History, Anatomy and Aesthetics of the Roman Letter Form*. New York: K. G. Saur, 1983.

Kepler, Johannes. *Mysterium Cosmographicum*. Tubingen: 1596. Augsburg: 1923, Munich/Berlin: 1936.
Levarie, Norma. *The Art and History of Books*. New York: Da Capo, 1982.

Lydenberg, Harry M. and Archer, John. *The Care and Repair of Books*, 4th rev. ed. Revised by John Alden, New York: Bowker, 1960.

Miner, Dorothy. *The History of Bookbinding, 525–1950 AD*. Baltimore: Walters Art Gallery, 1957.

Morison, Stanley. *Selected Essays on the History of Letter-Forms in Manuscript and Print*, 2 vols. Edited by D. McKitterick. Cambridge: Cambridge Univeristy Press, 1981.

Needham, Paul. *Twelve Centuries of Bookbindings, 400–1600*. New York: The Pierpont Morgan Library, 1979.

Prideaux, Sarah T. *Bookbinders and Their Craft*. New York: Charles Scribner's Sons, 1903.

Roberts, Matt and Etherington, Don. *Bookbinding and the Conservation of Books: A Dictionary of Descriptive Terminology*. Washington, D.C.: Preservation Office, The Library of Congress, 1982.

Young, Laura S. *Bookbinding and Conservation by Hand: A Working Guide*. New York: R. R. Bowker, 1981.

Zeier, Franz. *Paper Constructions: Two and Three Dimensional Forms for Artists, Architects and Designers*. Bern and Stuttgart: Paul Haupt, 1974.

SUPPLIERS

BOOKBINDING SPECIALISTS

Aiko's Art Materials Import, Inc.
714 North Wabash Avenue
Chicago, IL 60611
312-943-0745

Basic Crafts Company
1201 Broadway
New York, NY 10001
212-679-3516

Bookbinder's Warehouse, Inc.
45 Division Street
Keyport, NJ 07735
201-264-0306

BookMakers
2025 Eye Street NW, Room 502
Washington, DC 20006
202-296-6613

Conservation Resources
International, Inc.
1111 N. Royal Street
Alexandria, VA 22314
703-549-6610

Eighty Papers
510 Broome Street
New York, NY 10013
212-431-7720

Harcourt Bindery
51 Melcher Street
Boston, MA 02210
617-542-5893

Light Impressions Corp.
439 Monroe Avenue
Rochester, NY 14607
716-271-8960
800-828-6216 orders
800-828-9629 in NY

Pendragon
862 Fairmount Avenue
St. Paul, MN 55105
612-292-8666

Ernest Schaefer Inc.
731 Lehigh Avenue
Union, NJ 07083–7626
201-964-1280

TALAS
213 West 35 Street
New York, NY 10001–1996
212-736-7744

Twinrocker Papermill
R.R. #2
Brookston, IN 47923
317-563-3210

NATIONAL ART MATERIALS DEALERS

Dick Blick Company
P.O. Box 1267
Galesburg, IL 61401
800-447-8192 orders
800-373-7575 cust. service

Arthur Brown, Inc.
2 W 46 Street
New York, NY 10036
800-237-0619
718-779-6464 (in NY)

Sam Flax
39 W 19 Street
New York, NY 10011
212-620-3038

A.I. Friedman
24 W 45 Street
New York, NY 10036
212-243-9000

Pearl Paint
308 Canal Street
New York, NY 10013
212-431-7932

Flax, Pearl, and Blick have outlets in several parts of the U.S. Write for their catalogs to get complete listings.

IMPORTANT ADDRESSES

Center for Book Arts
626 Broadway
New York, NY 10012
212-460-9768

Guild of Bookworkers
521 Fifth Avenue
New York, NY 10175
212-757-6454

INDEX

A
Accordion: book, 18, 202, 203, 211; book, flexible, 212–214; book, with hard covers, 214–218; fold, 32, 211–216
Adhesive, 43–47, 55; applying, 48, 68, 213; bound book, 236; chart, 46. See also Gluing, Paste, PVA
Archimedean solids, 92

B
Board: bristol, 19, 22, 97, 99, 177, 248, 278; cover, 19, 99, 177, 265; covering, 53; edge, 62, 63, 67; illustration, 19, 99, 177, 248; mat; 19, 99, 278; museum, 177
Book-block, 217, 219–223, 263, 282; connecting case and, 279–282, 294, 298
Bone folder, 24, 27, 35, 51
Bottom margin, 98, 243, 261
Box: covering, 108, 117–118, 120–122, 169–170; edge, 108, 110, 112, 116, 120; folded, 105–107; hinged, 158–163; lining, 108–109, 118–120, 128, 163–164, 171; proportions, 105, 113; round, 164–174; with cover, 108–157
Bristol board, 19, 22, 97, 99, 177, 248, 278
Broadside, 32
Brochure, 240–249
Brushes, 27, 45, 47, 48

C
Cardboard, 16, 19, 22; construction, 66; gluing, 66–68; thickness, 16
Case, 277–287; connecting book-block and, 279–282, 294, 298, covering, 280–281
Chipboard, 19
Cloth, 23–24; gluing, 71–72, 162
Color: choice, 19, 23, 103, 109, 274, 285
Codex, 202, 225, 230–231
Collating, 243–244
Connection strips, 278–279, 293, 297
Corner, 62–63, 111, 115–116, 119; folding, 59; pinching, 56, 121, 163; proportions, 184; reinforcing, 188–191, 222
Counteracting tension, 53, 56, 108, 169
Cover: accordion book, 211, 213–216; boards, 19, 99, 177, 265; brochure, 240, 248–249; box, 41, 108, 113, 164–169, 172; fit, 113, 116, 120, 168; hinged, 123–124; palm-leaf book, 208–209; photograph album, 291, 293–296; placing, 50, 51, 277; reinforcing, 168, 169; thickness, 113
Creasing, 30, 35–36, 85, 177; guide, 36, 290, 293, 296. See also Folding
Cross fold, 32–33
Cube, 74–79, 130–131
Cut: beveled, 103; corner, 62
Cutting, 38–41; signatures, 33–35, 233

D
Deckle edge, 42, 101, 262
Dodecahedron, 74, 82, 90
Double leaves, 237, 257, 296, 297
Drying, 53; stack, 55

E
Edge, 49, 56; board, 62, 63, 67; book, 209, 279; box, 108, 110, 112, 116, 120; colored, 209, 274; covering, 126; deckle, 42, 101, 262; portfolio, 176, 190, 194, 197, 200; ragged, 39; reinforcing, 114–116, 125; trimming, 213, 215, 242, 243, 262
Endpapers, 56, 214, 220, 239, 258
Expandable bottom fold, 226

F
Flaps, 68, 168, 179, 194, 259; covering, 196; portfolio, 194–201
Flattening paper, 21
Fold, 31–35, 59; accordion, 32, 211–216; cross, 32; expandable bottom, 226; gate, 32; gluing, 87; mixed, 33; parallel, 32
Folding, 12, 30–36, 56, 58; corners, 59; mats, 99; shims, 297.
See also Creasing

G
Gauze, 24, 252
Gluing, 43–54; book cloth, 71–72; cardboard, 66–68; fold, 87; indirect, 49, 70–71; large formats, 51, 61–64; paper, 65–66; series of sheets, 73–74; spine, 238, 247–249, 271–273; tissue paper, 69–70; unsized cloth, 72–73. See also Adhesive; Paste; PVA
Grain direction, 13–15, 23, 33, 38, 167

H
Hardcover: book, 236, 250–251
Headband, 24, 251, 275, 276
Hinge, 37; in box, 158–163; in cover, 123–124; in mat, 97, 100–102; in portfolio, 175, 177, 196, 198, 200; in side, 124–126; material, 24, 199, 220; reinforcing, 177

I
Icosahedron, 74, 78, 80, 82, 91, 92
Illustration board, 19, 99, 177, 248
Inserting pictures, 257–258
Inside margin, 261

K

Kettle stitch, 263, 266–269. See also Sewing; Stitches
Knives, 24, 25, 27, 38–41
Knots, 210, 234, 235, 246, 263, 267–269. See also Sewing; Stitches

L

Labels, 70, 211, 232, 285–287
Lining, 64, 70; box, 108, 109, 118–120, 128, 163–164, 171; photograph album, 291–292, 294–295, 299; portfolio, 187–188, 190, 196, 216–217; strips, 24
Lip, 172–174
Lithography stones, 27, 54.
Long grain, 14
Loose-leaf book, 175, 203, 225–229

M

Margins, 98, 261
Mat, 96–103; board, 19, 99, 278; covering, 47–54, 103; folding, 99; hinges in, 97, 100–102; mounting in, 100–101; proportions, 97–98; uncovered two-piece, 101–102
Mi-teintes paper, 177
Moisture, 53–55, 67, 217, 229, 283, 295
Mounting: cardboard on cardboard, 66–68; drawing in mat, 100–101; paper on cardboard, 47–54, 56, 61–65; paper on paper, 65–66
Museum board, 177

N

Net, 76, 84, 88–92
Newsprint, 22, 48, 183, 190
Notebook, 233

O

Octahedron, 74, 78, 82, 84, 89, 92–95
Octavo, 34
Outside margin, 261
Overlap: at corners, 118, 121–122, 128, 163; at edges, 56, 118, 121, 163, 174, 192

P

Palm-leaf book, 202, 203, 207–210
Pamphlet, 232–235
Paper, 13–22; decorated, 7, 19; formats, 22; handling, 20; handmade, 19, 102, 214; Mi-teintes, 177; models, 80–95; samples, 17–20; stretching, 68–69; weight, 16,22
Papyrus, 202, 204, 230
Parallel folds, 32
Partitions, 127
Paste, 43–54; down, 283. See also Adhesive; Gluing; PVA
Photograph album, 288–300; adhesive-bound with cover, 293–295; lining, 291–292, 294–295, 299; sewn with cover, 296–300; side-sewn, 289–292
Platonic solids, 74, 75, 78–79
Polyhedra, 79, 91–93
Portfolio: covering, 187–190, 192; cloth, 192; cloth spine, 181–183, 187–190; creased and scored, 177–179; edges, 176, 190, 194, 197, 200; hinges in, 175, 177, 196, 198, 200; lining, 187–188, 190, 196, 216–217; parts, 176; proportions, 197; with expandable bottom fold, 226–229; with flaps, 194–201; with ties, 180
Proportions: book cover, 280; box, 105, 113; corner, 184; mat, 97–98; portfolio, 197
Punching holes, 210, 222, 223, 290
PVA, 43–46, 66–67, 72. See also Adhesive; Gluing; Paste

R

Reinforcing: corner, 188–191, 222; cover, 168, 169; edge, 114–116, 125; hinge, 177; signatures, 256; spine, 276; strips, 254, 255
Rounding: right and wrong, 265, 273; spine, 264, 272–273

S

Sandpaper file, 25, 27, 66
Scoring, 37
Scroll, 204–206
Sewing, 244–246; side, 219, 224; signatures, 244–246, 253, 266–270; tapes, 263. See also Kettle stitch; Knots; Stitches
Shims, 288, 289, 296; folding, 297
Short grain, 14
Side margin, 98, 261
Signatures, 34–35, 230, 233, 235, 240, 243; marks, 252–253; reinforcing, 256; sewing, 244–246, 253, 266–270; thickness, 34, 35, 235, 246, 258
Smoothing colored edges, 274
Softcover, 240, 277–278
Spine, 56, 176, 183, 184, 195, 198, 236–238, 264–265; covering, 275–276, 278–281; flat, 250, 272–273, 278, 294, 296; flexible, 181–182; gluing, 247, 271–273; labelling, 285–287; reinforcing, 276; rigid, 159; rounding, 264, 272–273
Stitches, 244, 249, 265, 267, 292; distribution, 220–224; kettle, 263, 266–269; variation, 219–224. See also Knots; Sewing
Stretching paper, 68–69
Swell, 245, 264–265; adjusting, 264

T

Tapes: sewing, 263
Tearing, 38
Template: for applying adhesive, 68; for punching holes, 210; for trimming signatures, 215, 242
Tension, 46, 66, 246, 256; counteracting, 53, 56, 108, 169
Tetrahedron, 74, 76, 78, 80, 82, 84–87
Thread, 27, 219, 223, 246, 253, 263–269, and swell, 245, 264–265, 277
Ties, 186
Titling, 284
Tools, 24–27
Top margin, 98, 261
Triangle, 25, 26
Trimming: book, 221, 261–262; brochure, 242; edges, 213, 215, 242–243, 262; sheet, 42

W

Warping, 29, 55, 61, 64, 73
Weighting, 53–54, 233, 259
Weights, 27
Work space, 28–29